A "Hands On" approach to teaching...

Algebra

Linda Sue Brisby

Andy Heidemann

Natalie Hernandez

Jeanette Lenger

Ron Long

Petti Pfau

Scott Purdy

Sharon Rodgers

HANDS ON, INC.

SOLVANG, CALIFORNIA

Layout and Graphics: Scott Purdy
Primary Layout: Linda Sue Brisby and Jeanette Lenger
Illustrators: Petti Pfau, Sharon Rodgers, Suzi Matthies
Cover Art: Petti Pfau

Copyright © 1990 by HANDS ON, INC. All rights reserved. Printed in the United States of America.

Limited Reproduction Permission: The authors and publishers hereby grant permission to the teacher who purchases this book, or the teacher for whom this book is purchased, to reproduce up to 100 copies of any part of this book for use with his or her students. Any other duplication is prohibited.

Order Number: HO 105
ISBN 0-927726-03-3

HANDS ON, INC.
2121 Rebild Drive
Solvang, CA 93463

Introduction

"How do you do algebra in kindergarten?" "Algebra is a high school subject," "Elementary kids don't need algebra," and "I never understood it in school, and now I'm supposed to teach it to primary school children?" These are the typical responses of teachers when confronted with the idea of teaching algebra in elementary grades.

Nevertheless, we charged forward, convinced that it could be done. Now that this book is complete, we are more convinced than ever that algebra CAN and MUST be a part of every child's math education. And interestingly, most teachers are teaching a lot more algebra than they think.

What is algebra for kindergarten through junior high age children?
   It is solving for unknowns -- something that children begin doing in kindergarten with exercises such as $1 + \square = 3$.
   It is identifying relationships between numbers, which we begin teaching with the use of number lines and activities such as $5 < 6 < 7$.
   It is identifying true and false math statements
   It is interpreting and processing story problems along with identifying unnecessary or missing information.
   It is understanding that $3 + 5$ and $5 + 3$ are the same, as are $8 \times 4 \times 5$ and $5 \times 4 \times 8$.
   It is finding solution sets and understanding that there can be more than one right answer in a math problem.
   It is understanding that formulas are equations and equations are a means of representing a relationship in value.

We also found that algebra lends itself to many cross-curricular opportunities and have provided many integrated lessons involving art, literature, science, and social studies. In addition, we continue with our emphasis on cooperative grouping.

Activities are organized around a TASK ANALYSIS of the skills which children need to understand Algebra. Lessons are organized from simple to complex within each task analysis item, and each lesson has a list of materials, recommended classroom organization, and a basic explanation of the lesson format. We have also included extensions in many lessons which will allow for elaboration for high ability students.

What we have provided is over 150 organized, concise, activity oriented lessons for teaching algebra. We have included kindergarten through ninth grade lessons so any elementary teacher will have access to remedial and enrichment lessons in one convenient source. A second benefit of this format is that teachers can see the whole spectrum of the material that elementary age children should be learning.

As with our other books, Hands on Statistics, Probability and Graphing, Hands on Measurement, Hands On Logic, and Hands on Geometry, this book is written by eight teachers in Solvang, California. We are writing this math series while we continue to work in the classroom. All of our books are designed to align with both the California State Mathematics Framework and the National Council of Teachers of Mathematics (NCTM) Curriculum Standards.

Algebra
Task Analysis

Variables	1. Identifies, uses and creates sets of variables and/or numbers used to replace unknowns.	Primary
Operation Symbols	2. Identifies and uses + and – symbols to show relationships between numbers.	Primary
<, >, =	3. Identifies and uses <, >, and = symbols to show relationships between numbers.	Primary
True/False	4. Identifies and creates true/false number sentences.	Primary
Open Sentences	5. Identifies and creates open number sentences.	Primary
Variables	6. Identifies and creates missing one digit addends, minuends, and subtrahends in number sentences.	Primary
Word Problems	7. Identifies, uses and creates equations to represent addition and subtraction word problems to show missing information.	Primary
Variables	8. Recognizes and identifies variables and equations.	Middle
<, >, =	9. Identifies basic level symbols <, >, and =.	Middle
Variables	10. Identifies sets and/or numbers used to replace variables.	Middle
True/False	11. Identifies true or false statements or number sentences.	Middle
Complex Symbols	12. Identifies complex symbols: parentheses, brackets, intersections, etc.	Middle
Properties	13. Identifies and uses the commutative property.	Middle
Properties	14. Identifies and uses the associative property.	Middle
Properties	15. Identifies aand uses the distributive property.	Middle/Upper
Word Problems	16. Identifies extraneous or missing information in word problems.	Middle/Upper

Hands On, Inc
2121 Rebild Drive
Solvang, CA 93463

Algebra

Open/Closed	17. Identifies open and closed number sentences.	Middle/Upper
Operation Order	18. Applies order to operations: multiplication, division, addition, subtraction.	Middle/Upper
Positive/Negative	19. Uses number lines to solve equations with positive and negative numbers.	Middle/Upper
Formulas	20. Uses number sentences or formulas to represent a value.	Middle/Upper
Variables	21. Uses various methods to solve for variables in simple equations.	Middle/Upper
Variables	22. Uses various methods to solve for variables in complex equations.	Upper
Word Problems	23. Uses equations to solve problems with decimals or fractions.	Upper
Linear Equations	24. Uses linear equations to represent relationships.	Upper
Complex Symbols	25. Creates equations employing parentheses and brackets.	Upper
<, >, =	26. Creates equations and number sentences including symbols <, >, =.	Upper
Variables	27. Creates equations using variables.	Upper
Word Problems	28. Creates equations from word problems.	Upper
Number Patterns	29. Creates equations which reflect number patterns.	Upper
Venn Diagrams	30. Creates solutions sets for equations using Venn Diagrams, unions, intersections, etc.	Upper
Word Problems	31. Creates word problems based on given equations.	Upper
Tables/Graphs	32. Creates equations based on information found in tables or graphs.	Upper
Simplifying	33. Creates a simplified expression of an equation.	Upper

About the Authors...

Linda Sue Brisby	M.A. – Indiana University; B.A. – Goshen College; Reading Specialist – University of Michigan; 19 years of teaching experience in primary grade levels.
Andy Heidemann	B.A. – Ohio Wesleyan University; 6 years of teaching experience in upper elementary and junior high.
Natalie Hernandez	B.A. – University of Illinois; 22 years of teaching experience in various middle grade levels.
Jeanette Lenger	B.S. – Cal Poly State University, San Luis Obispo; 23 years of teaching experience in primary grade levels.
Ron Long	M.A. – Cal Poly State University, San Luis Obispo; B.S. – Long Beach State University; 21 years of teaching experience at the junior high level.
Petti Pfau	M.A. – Cal Poly State University, San Luis Obispo; B.A. – San Francisco State University; 17 years of teaching in middle elementary grades and in special education.
Scott Purdy	M.A. – Western State College, Colorado; B.A. – UCLA: 17 years of teaching/administration from middle elementary through high school level.
Sharon Rodgers	B.S. – Southern Oregon College; 26 years of teaching experience at all elementary through junior high levels.

Algebra

Table of Contents

Fall Activities – Primary
    Add Me – Subtract Me …………………………………………………………… 1
    I Must Be True …………………………………………………………………… 3
    Apple = Apple …………………………………………………………………… 7
    Greater Than What ……………………………………………………………… 9
    Football Frolic …………………………………………………………………… 11
    Number Line Find ……………………………………………………………… 13
    My Changing Train ……………………………………………………………… 15
    True Blue ………………………………………………………………………… 17
    One Diamond = Two Triangles ………………………………………………… 19
    Beary True ……………………………………………………………………… 21
    Party Time ……………………………………………………………………… 23
    How Nutty? ……………………………………………………………………… 25

Winter Activities – Primary
    Kissing Elves …………………………………………………………………… 29
    So Many Children ……………………………………………………………… 31
    What Will You Find In Your Stocking? ………………………………………… 33
    Bears Love Honey ……………………………………………………………… 35
    Crackered-up …………………………………………………………………… 39
    Rebus Bear Walk ……………………………………………………………… 41
    "100" Bags ……………………………………………………………………… 43
    My Silly Valentine ……………………………………………………………… 45
    How Many Coins Do Your Crayons Weigh? …………………………………… 49

Spring Activities – Primary
    Windy Kites ……………………………………………………………………… 51
    Peanuttiest ……………………………………………………………………… 53
    Tater Time ……………………………………………………………………… 55
    Rain Rain Go Away …………………………………………………………… 59
    How Heavy Is Your Egg? ……………………………………………………… 61
    Seedy Business ………………………………………………………………… 65
    What's True For You? ………………………………………………………… 67
    What Do You Like Best In Language Arts? ………………………………… 69
    I Scream, You Scream ………………………………………………………… 73

Hands On, Inc
2121 Rebild Drive
Solvang, CA 93463

Algebra

8. Recognizes and identifies variables and equations.
    - Equalities in Advertising — 75
    - Cheeseburger, Cheeseburger — 76
    - Playing a Part — 77
    - Red and Yellow Make... Variables — 78
9. Identifies basic level symbols <, >, and =.
    - Wheeling and Dealing — 79
    - Know it Alls — 80
    - Picture This — 81
    - Call Your Own Number — 82
10. Identifies sets and/or numbers used to replace variables.
    - Rollin' Around — 83
    - Cryptograms — 84
    - Dollars and Sense — 85
    - The Puzzling U.S. — 86
11. Identifies true or false statements or number sentences.
    - What You See Is What You Get — 87
    - State the Whole Truth — 88
    - Stand Up for Sentencing — 89
    - The Domino Effect — 90
12. Identifies complex symbols: parentheses, brackets, intersections, etc.
    - What's Your Algebraic Sign? — 91
    - Serve Up a Symbol — 92
    - Sentence Savvy — 93
    - Have You Got a Match? — 94
13. Identifies and uses the commutative property.
    - It's All Bean Said Before — 95
    - Element..ary My Dear Bean — 96
    - Hip, Hip Array — 97
    - Staring at Stairs — 98
14. Identifies and uses the associative property.
    - Round and Round You Go — 99
    - What Goes Where, When? — 100
    - Enjoying the Easy Times — 101
    - Multiplying Darts — 102
15. Identifies and uses the distributive property.
    - What's Cookin' — 103
    - Here's a Tip for You — 104
    - Distributing Discounts — 105
    - Mental Math — 106
16. Identifies extraneous or missing information in word problems.
    - Word Wary — 107
    - Finding the Rhyme and Reason — 108
    - Checking the Stats — 109
    - Extra Parts — 110

Hands On, Inc
2121 Rebild Drive
Solvang, CA 93463

Algebra

17. Identifies open and closed number sentences.
    To Loop or Not to Loop — 111
    An Indefinite Answer — 112
    Missing Jigsaw — 113
    Closing Out the Game — 114

18. Applies order to operations: multiplication, division, addition, subtraction.
    Placing an Order — 115
    I'll Even Give You the Answer — 116
    Around and Around — 117
    Valuable Letters — 118

19. Uses number lines to solve equations with positive and negative numbers.
    AD/BC — 119
    Adding Nothing? — 120
    Double Negatives — 121
    Submarine Fuelishness — 122

20. Uses number sentences or formulas to represent a value.
    Driving Off Into the Sunset — 123
    Shoebox Algebra — 124
    You Can Even Add Odd — 125
    Name That Tune — 126

21. Uses various methods to solve for variables in simple equations.
    Equatiocard — 127
    Does Your Body Count? — 128
    Finding a Weigh — 129
    A Good Sense of Spell — 130

22. Uses various methods to solve for variables in complex equations.
    Call for an Operation — 131
    An Eye for an Eye and a Bean for a Bean — 132
    Do Unto Others — 133

23. Uses equations to solve problems with decimals or fractions.
    Have it Your Way — 135
    All Mixed Up — 136
    Tricky Taxes — 137
    Body Weight — 138

24. Uses linear equations to represent relationships.
    A Penny a Pound — 139
    Linear Circles? — 140
    Getting the Drop on Linear Equations — 141
    I've Gotta Split — 142

25. Creates equations employing parentheses and brackets.
    Greater Than a Book — 143
    A Well Balanced Diet — 144
    Don't Lose Your Cookies — 145
    Anything but Chili — 146

Hands On, Inc
2121 Rebild Drive
Solvang, CA 93463

Algebra

26. Creates equations and number sentences including symbols <, >, =.
    - On a Roll — 147
    - Why? — 148
27. Creates equations using variables.
    - Wheeler Dealer — 149
    - Name Games — 150
    - Rebus Variables — 151
    - A Group "Outing" — 152
28. Creates equations from word problems.
    - Word Problem Lotto — 153
    - Dear Abby — 154
    - Roamin' Around the Numerals — 155
    - Now That's a Tall Tale — 156
29. Creates equations which reflect number patterns.
    - Cubist Math — 157
    - Time for My Formula — 158
    - What's it Worth to You — 159
    - Watch Out for the Algebug — 160
30. Creates solutions sets for equations using Venn Diagrams, unions, intersections, etc.
    - Ring Around the Answer — 161
    - Relating the Unrelated — 162
    - Varied Ads = Variables — 163
    - Equating the Intersections — 164
31. Creates word problems based on given equations.
    - It Adds Up to My Family — 165
    - Records With a Club — 166
    - Equation Collage — 167
    - The Whys of Ups and Downs — 168
32. Creates equations based on information found in tables or graphs.
    - Making Converts — 169
    - Sometimes You Win — 170
    - A Yen For Formulas — 171
    - Three Coins in a Pocket — 172
33. Creates a simplified expression of an equation.
    - Bringing a Monopoly Down to Size — 173
    - Bean Counters Take Heart — 174
    - Have You Got Five Fizzles for a Klatch? — 175
    - Thanks, but No Thanks — 176

Appendices — 177

Algebra

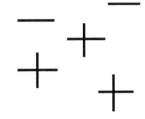

Add Me – Subtract Me
September
Grade Level: Primary

TASK ANALYSIS: 2 – Identifies, uses, and creates + and – symbols to show relationship between numbers

MATERIALS: Gobs of tagboard strips 2" x 12", felt pens, large sheet of butcher paper, sandpaper, clay, sand/salt, pudding/fingerpaint, paint, c-rods, toothpicks, unifix cubes, tiles, straws or any other junk that is available to you

ORGANIZATION: Whole group presentation
Cooperative groups of 2-4 children to rotate through stations
K-3

PROCEDURE:
- Teacher presents plus symbol to entire group using the tagboard strips.
- Do each of the following:
- Children "draw" symbol in the air as they say symbol's name.
- Children "draw" symbol on each other's back as they say the symbol's name.
- Children may volunteer to draw the symbol on the butcher paper with the felt pen.
- Children proceed through station rotation creating the plus symbol.
- Teacher presents minus symbol.
- Children repeat the above procedure.

- Note to teacher: symbols may be taught in any order.
- Station descriptions:
- Sandpaper: enough sandpaper strips should be available so each child may create each symbol, constuction paper and glue.
- Each child creates and mounts his symbols on the construction paper. Children then "trace" each symbol.

Algebra

Children may do this with eyes opened and closed.

Clay: mats on which to roll the clay, different colors of clay, crayons, and white paper.
Children roll out clay "snakes" and form symbols.
Children record with crayons on the white paper the symbol they created.

Salt and/or sand: salt/sand, pie tins or 9" x 11" shallow box lids
Children create symbol with fingers in the sand.

Pudding/fingerpaint: paper, fingerpaint/pudding, plastic to cover work area, quarter-cup measure
Children may take one measure of mixture and create symbol.
Paint: paint, paper, brushes
Children create symbol.

C-rods: cuisinaire rods, crayons, paper
Children create symbol with the rods.
Children record with crayons on paper the symbol created.

Toothpicks: flat toothpicks, glue, paper
Children create symbol with toothpicks.
Children record by gluing the toothpick symbols to the paper.

Unifix cubes: unifix cubes, colored one-inch paper squares, glue, paper
Children create symbol with unifix cubes.
Children record by re-creating the symbol with the paper squares.
Children glue squares to paper.

Straws: straws, paper, glue
Children create symbol with straws.
Children record by gluing the straws to the paper.

Algebra

I Must Be True
September
Grade Level: Primary

TASK ANALYSIS: 1 – Identifies, uses, and creates sets of variables and/or numbers used to replace unknowns

MATERIALS: Butcher paper wall chart, felt pens, floor equation mat, apples, xeroxed copy of record sheet (for 2nd-3rd: 1 per child), pencils, twenty 2" x 12" tagboard strips to form algebraic symbols

ORGANIZATION: Whole group activity for presentation
K-1 continues as whole group activity
2-3 continues in cooperative learning groups

PROCEDURE:
- Make wall chart with headings and columns.
- Place equation mat on floor.

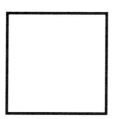 

- Present the symbols of + and =.
- Have children place on equation mat.
- Teacher presents the "givens" of the equation. (+ 1 = 5)
- Ask the questions: "How many apples can go in the box to make this a true number sentence?" "What other ways may the apples be placed to make this a true number sentence?"
- Note to teacher: At this point you want to help the children "see" the available number combinations that create a set of "true" number sentences.

Algebra

Record each true set on the wall chart.
Continue until children have exhausted all possible sets.
Discuss and interpret data.

□	+	?	=	?
🍎🍎🍎🍎	+	🍎	=	🍎🍎🍎🍎🍎
🍎🍎	+	🍎🍎🍎	=	🍎🍎🍎🍎🍎
	+		=	

2nd-3rd grades will continue in cooperative groups.
Record the information from the wall chart onto the recording sheet using numerals.
As whole class compare and discuss cooperative group record sheets.

I Must Be True
Recording Sheet

▭	+	?	=	?

Algebra

Apple = Apple
Recording Sheet

Estimate	Actual	> < =

Algebra

Apple = Apple
September
Grade Level: Primary

TASK ANALYSIS: 2 – Identifies, uses, and creates +, and – symbols to show relationship between numbers
3 – Identifies, uses, and creates <, >, and = symbols to show relationship between numbers

MATERIALS: Balance scales for each cooperative group, tub of unifix cubes, tub of teddy bear counters, or tub of pattern blocks for each cooperative group, one apple per cooperative group, xerox copy of record sheet (one per group), pencils

ORGANIZATION: Whole group activity for presentation
Cooperative learning groups
K–3

PROCEDURE:
- Class decides which arbitrary unit of measure will be used by each cooperative group. This must be the constant for comparing the equations.
- Model weighing an apple in the balance scales with the chosen unit of measure.
- Form cooperative groups.
- Each cooperative group places apple in balance scale.
- Group then estimates the number of tub items needed to balance the scale.
- Record estimate.
- Each group adds items to make scale balance.
- Count and record the actual number of items needed to balance.
- Each group gives number sentence using the equation created by comparing the estimate to the actual number of items.
- Each group records their own equation.
- Two cooperative groups work together to compare and interpret data.

Algebra

Each child creates and mounts his symbols on the construction paper. Children then "trace" each symbol.
Children may do this with eyes opened and closed.

Clay: mats on which to roll the clay, different colors of clay, crayons, and white paper.
Children roll out clay "snakes" and form symbols.
Children record with crayons on the white paper the symbol they created.

Salt and/or sand: salt/sand, pie tins or 9"x11" shallow box lids
Children create symbol with fingers in the sand.

Pudding/fingerpaint: paper, fingerpaint/pudding, plastic to cover work area, quarter-cup measure
Children may take one measure of mixture and create symbol.

Paint: paint, paper brushes
Children create symbol.

C-rods: Cuisinaire rods, crayons, paper
Children create symbol with the rods.
Children record with crayons on paper the symbol created.

Toothpicks: flat toothpicks, glue, paper
Children create symbol with toothpicks.
Children record by gluing the toothpick symbols to the paper.

Unifix cubes: unifix cubes, colored one-inch paper squares, glue, paper
Children create symbol with unifix cubes.
Children record by re-creating the symbol with the paper squares.
Children glue squares to paper.

Straws: straws, paper, glue
Children create symbol with straws.
Children record by gluing the straws to the paper.

Algebra

Greater than What?
September
Grade Level: Primary

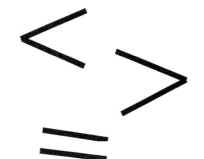

TASK ANALYSIS: 3 – Identifies, uses, and creates <, >, and = symbols to show relationship between numbers

MATERIALS: Gobs of tagboard strips 2" x 12", felt pens, large sheet of butcher paper, sandpaper, clay, sand/salt, pudding/fingerpaint, paint, c-rods, toothpicks, unifix cubes, tiles, straws or any other junk that is available to you.

ORGANIZATION: Whole group presentation
Cooperative groups of 2-4 children to rotate through stations
K-3

PROCEDURE:
- Teacher presents greater than symbol to entire group using tagboard strips.
- Do each of the following:
- Children "draw" symbol in the air as they say symbols name.
- Children "draw" symbol on each others back as they say the symbols name.
- Children may volunteer to draw the symbol on the butcher paper with the felt pen.
- Children proceed through station rotation creating the greater than symbol.
- Teacher presents the less than symbol.
- Children repeat the above procedure
- Teacher presents the equal to symbol.
- Children repeat the above procedure.
- Note to teacher: symbols may be taught in any order.

- Station descriptions:
- Sandpaper: enough sandpaper strips should be available so each child may create each symbol, construction paper and glue.

Algebra

Football Frolic
October

Grade Level: Primary

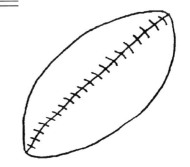

TASK ANALYSIS: 3 – Identifies, uses, and creates <, >, and = symbols to show relationship between numbers
4 – Identifies, uses, and creates true/false number sentences
5 – Identifies, uses, and creates open number sentences

MATERIALS: Yarn or hula hoops to make large circles on the floor, real footballs, butcher paper, felt markers, tagboard symbols of =, <, and >

ORGANIZATION: Whole group activity for presentation
This is a great activity for small groups or individual children to do once they are familiar with the procedure.
K–3

PROCEDURE:
- Make butcher paper graph divided in half with headings (true/false) and three tagboard signs showing =, <, and >.
- Attach the graph to the wall.
- Place two circles on the floor with the equal sign between them.

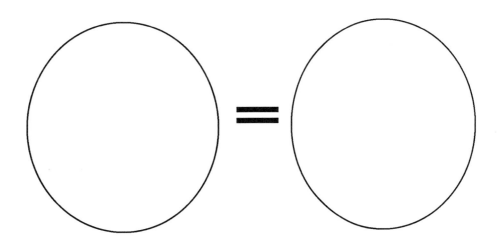

Algebra

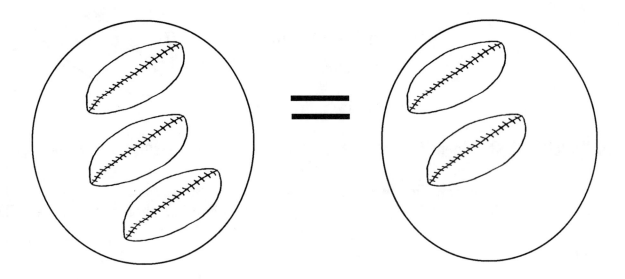

Teacher places 3 footballs in first circle, and 2 footballs in second circle.
Ask the question: "Are there the same number of footballs in the first circle as in the second circle?"
      "Is 3 the same as 2?"
      "Does the first circle equal the second circle?"
      "Does 3 = 2?"
Then the teacher records under the "false" heading 3 (pictures of footballs = 2 (pictures of footballs).
Children use their footballs to create true/false equations on the floor circles. The teacher will record the children's findings on the wall graph.
Once the children are comfortable with equalities, the next step will be to repeat this procedure using < or > to create true/false number sentences.

Extensions: You may use autumn leaves or nuts to do this lesson.

Algebra

Number Line Find
October
Grade Level: Primary

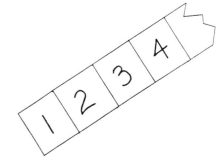

TASK ANALYSIS: 1 – Identifies, uses and creates sets of variables and/or numbers used to replace unknowns
2 – Identifies, uses, and creates + and symbols to show relationship between numbers
3 – Identifies, uses, and creates <, >, and = symbols to show relationship between numbers
4 – Identifies, uses, and creates true/false number sentences
5 – Identifies, uses, and creates open number sentences
6 – Identifies, uses and creates missing one-digit addends and subtrahends in number sentences

MATERIALS: Number line displayed in room (example on next page), or 2nd and 3rd graders may use individual number lines

ORGANIZATION: Whole class activity for presentation
K-1 continue as whole class activity
2-3 cooperative groups

PROCEDURE: – Using the number line in your classroom, the teacher first models the given number:
  – "I am a number ."
  – "I have (1, 2, or 3, etc.) digits."
  – "You will find me between and ____ ."
  – "I am greater than ____ ."
  – "I am less than ____ ."
  – "What number am I?"
  – Children may ask:
  – "Are you > _?_"
  – "Are you < _?_"
  – "Are you = _?_"

Algebra

"Are you an even number?"
"Are you an odd number?"
Choose a "Number Child" to pick a number and do the above procedure.
The "Number Child" answers in sentences using <, >, etc.
Children are free to discuss the questions and answers as they move through the possible solutions.
Child with the final solution becomes the new "Number Child" and repeats the procedure.

Extension: We find this to be a tremendous sponge activity with the children throughout the year.

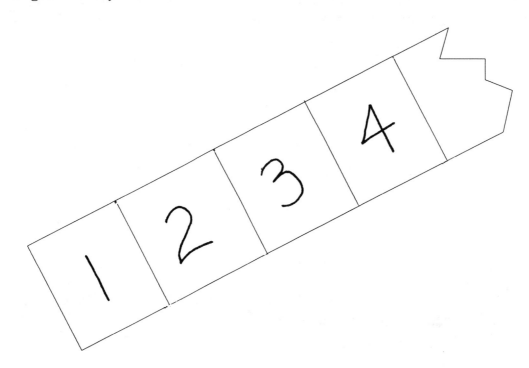

Algebra

My Changing Train
October

Grade Level: Primary

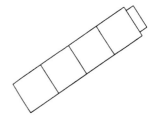

TASK ANALYSIS: 1 – Identifies, uses, and creates sets of variables and/or numbers used to replace unknowns

MATERIALS: Unifix cubes, record sheets, symbols

ORGANIZATION: Whole group activity for presentation
Cooperative groups of 2-4 children
K-3

PROCEDURE: – Model several unifix trains

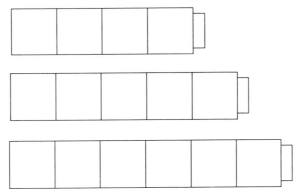

- Children work in groups of 2-4 children.
- Record several of their "variables" (combinations) on a record sheet.

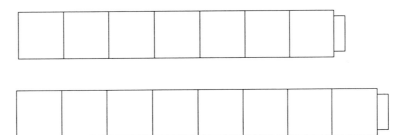

My Changing Train
Recording Sheet

Algebra

True Blue
October

Grade Level: Primary

TASK ANALYSIS: 3 – Identifies, uses, and creates <, >, and = symbols to show relationship between numbers
4 – Identifies, uses, and creates true/false number sentences.

MATERIALS: Yarn or hula hoops to make large circles on the floor, unifix cubes, apples, butcher paper, felt markers, tagboard signs of =, <, and >

ORGANIZATION: Whole group activity for presentation
This is a great activity for small groups or individual children to do once they are familiar with the procedure.
K-3

PROCEDURE:
- Make butcher paper graph divided in half with headings (true/false) and three tagboard signs showing =, < and >.
- Hang graph.
- Place 2 circles on the floor with the equal sign between them.

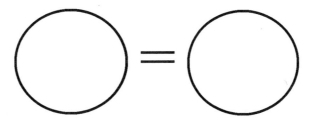

- Teacher places 3 apples in first circle, and 2 apples in second circle.
- Ask the question: "Are there the same number of apples in the first circle as in the second circle?" "Is 3 the same as 2?" "Does the first circle equal the second circle?" "Does 3 = 2?"
- Then teacher records under the "false" heading 3 (pictures of apples) = 2 (pictures of apples).

Algebra

Now the children use their apples to create true/false equations in the floor circles.
Teacher will record children's findings.
Once the children are comfortable with equalities, the next step will be to repeat this procedure using < or > to create true/false number sentences.

TRUE	FALSE
	🍎🍎🍎 = 🍎
🍎🍎 = 🍎🍎	🍎 = 🍎🍎
	🍎 = 🍎🍎🍎

Algebra

One Diamond = Two Triangles
November
Grade Level: Primary

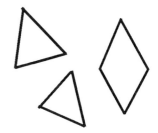

TASK ANALYSIS: 1 – Identifies, uses, and creates sets of variables and/or numbers used to replace unknowns
2 – Identifies, uses, and creates + and – symbols to show relationship between numbers
3 – Identifies, uses, and creates <, >, and = symbols to show relationship between numbers
4 – Identifies, uses, and creates true/false number sentences

MATERIALS: Pattern blocks

ORGANIZATION: Cooperative groups of 2–4 children
K–3

PROCEDURE: – Model pattern blocks of various shapes stacked.

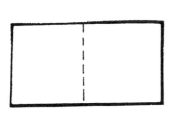 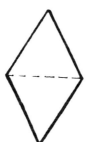 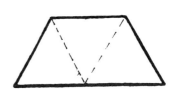

- Explore with your blocks to discover what sets of blocks you believe are =, <, >.
- Children explore for 10–20 minutes.
- Gather as a whole group to share discoveries.

Algebra

"We discovered 3 △ = 1 ⌂ ."

"We discovered 2 △ = 1 ◇ ."

"We discovered 6 △ = 1 ⬡ ."

After discoveries are shared and recorded on large wall chart or floor chart ask, "What may you add or subtract to make it an untrue number sentence.
    i.e. 3 ___ = 1 ___   true

        2 ___ = 1 ___   false

Algebra

Beary True
November
Grade Level: Primary

TASK ANALYSIS: 2   Identifies, uses and creates + and − symbols to show relationship between numbers
3 − Identifies, uses and creates <, >, and = symbols to show relationship between numbers
4 − Identifies, uses and creates true/false number sentences
6 − Identifies, uses and creates missing one-digit addends and subtrahends in number sentences

MATERIALS: Yarn to make large circles on the floor, felt markers, butcher paper, children's stuffed bears, tagboard symbols

ORGANIZATION: Whole group activity
K−3

PROCEDURE:
- Make true/false butcher paper graph.
- Hang graph on the wall.
- Introduce tagboard signs of =, <, >.
- Place yarn circles on floor with equal sign:

 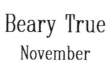 

- Teacher asks: "Put all the black bears in the first circle and all the brown bears in the second circle. Who can give a number sentence?"

Algebra

"Is that number sentence true or false?"
Record on chart.
"Who can use an algebraic symbol to make the sentence true?"
Record true equations on the chart.
Use the following bear questions to continue the discovery:
  Are there:
    more white bears than black bears?
    less brown bears than white bears?
  Do:
    White bears + black bears = brown bears?
    Brown bears + black bears > white bears?

Based upon the bears your students brought continue the questioning.
Record data.
Children discuss and interpret data.

Algebra

Party Time
November
Grade Level: Primary

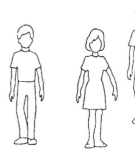

TASK ANALYSIS:
1 – Identifies, uses and creates sets of variables and/or numbers used to replace unknowns
2 – Identifies, uses and creates + and − symbols to show relationship between numbers
3 – Identifies, uses and creates <, >, and = symbols to show relationship between numbers
6 – Identifies, uses and creates missing one-digit addends and subtrahends in number sentences

MATERIALS: Small milk cartons (or other small containers), unifix cubes, butcher paper chart, marker

ORGANIZATION: Whole group activity
K-3

PROCEDURE:
- Children plan a birthday celebration.
- They invite Linda's class
- They know they will need 24 party treats.
- They need to know how many boy treats and girl treats, there are 5 girls, but how many boy treats are needed?
- On the floor, set milk cartons (or any small container) to represent each child.

Algebra

Decide which object is the girl treat and the boy treat.
Green unifix cube = girl; burgundy unifix cube = boy.
They know Linda's class has 5 girls.
Children place one green unifix cube in each of 5 cartons.
Show equation on butcher paper chart.

$$5 + \square = 24$$
$$10 + \square = 24$$
$$12 + \square = 24$$

Children place one burgundy unifix in each of the remaining containers.
Children discuss the information they've created. (They know there are 24 children, 24 cartons and each carton has a birthday treat (unifix cube). (They know there are 5 girls [5 green unifix cubes]).
Ask "How can you find the number of boy treats needed?"
Have children count burgundy unifix cubes.
Children fill in unknown box on the butcher paper chart.
Read equation.
Discuss how it would be different if there were 10 girls, 12 girls, etc. coming to the party.
Discuss which is $<$, $>$, or $=$ to.

Algebra

How Nutty?
November
Grade Level: Primary

TASK ANALYSIS: 2 – Identifies, uses and creates + and – symbols to show relationship between numbers
3 – Identifies, uses and creates <, >, and = symbols to show relationship between numbers

MATERIALS: Balance scales for each cooperative group, tub of unifix cubes, tub of unifix cubes, or tub of pattern blocks for each cooperative group, 2 cups of nuts per cooperative group (walnuts or peanuts in the shell work the best), xerox copy of record sheet (one per group), pencils

ORGANIZATION: Whole group activity for presentation
Cooperative learning groups
K-3

PROCEDURE:
- Class decides which arbitrary unit of measure will be used by each cooperative group. This is the constant for comparing the equations.
- Teacher models weighing the nuts in the balance scales with the chosen unit of measure.
- Form cooperative groups.
- Each cooperative group places nuts in balance scale.
- Estimates the number of tub items needed to balance the scales.
- Record estimate.
- Each group adds items to make scale balance.
- Count and record the actual number of items needed to balance.

Algebra

Each group gives number sentence using the equation created by comparing the estimate to the actual number of items.
   Example:
      "I estimated 10 unifix cubes."
      "I needed 14 unifix cubes."
      "I estimated 4 cubes < I needed."
Each group records their own equation.
Then have two cooperative groups work together to compare and interpret each group's data.

Extensions: Use apples or toy footballs instead of nuts.

Algebra

How Nutty?
Recording Sheet

Estimate Actual > < =

Algebra

Kissing Elves
December
Grade Level: Primary

TASK ANALYSIS: 1 – Identifies, uses, and creates sets of variables and/or numbers used to replace unknowns
4 – Identifies, uses and creates true/false number sentences
5 – Identifies, uses, and creates open number sentences

MATERIALS: Butcher paper, markers, Hershey kisses (2 per child), book or record of <u>The Elves and the Shoemaker</u>, a note from the elves

ORGANIZATION: Whole class activity
K–3
2nd and 3rd grades may do extension individually or in cooperative groups

PROCEDURE:
- Teacher writes note to the class from the "Elves" and attaches it to a bag of Hershey kisses (see example on next page).
- Read and discuss <u>The Elves and the Shoemaker</u>.
- Children brainstorm for what other things elves may bring.
- Children read the note left by the elves that tells them to find the bag of kisses.
- Children search for the bag of kisses (make it easy to find).
- Search for about 5 minutes.
- Children brainstorm how to find an answer to the elves' questions.
- Teacher reads the graph chart the elves left as a clue for the children to solve the problem.
- Children then each get two kisses (eat them of course!) and save the wrappers.
- Build the elf graph one child at a time.
- Children fill in the missing numbers on the chart.
- Children discuss how to make the sentences true or false.
- Discuss and interpret data.

Algebra

Extensions: Second and third grade children write letters to the elves explaining how to solve the dilemma.

First graders write sentences on long, lined story paper to tell the elves how to solve the problem.

Kindergartners dictate ideas to teacher so he/she may write them down for the elves.

Dear First Graders (use your grade level),

    We visited your classroom last night and really enjoyed the wonderful pictures and displays in your room.

    We would like to be in your school everyday. You are learning so many interesting things.

    We saw all your Math tubs and wondered if you could help us solve our problem. We left you a chart to show you all we know so far.

    Please let us know if you can help us solve it.

    Love,

    The Elves

PROBLEM

One elf came to the Christmas party last year and brought two kisses.
Two elves came and brought four kisses.
Three elves came and brought six kisses.
Four elves came and brought __ kisses.
Five elves came and brought __ kisses.
Six elves came and brought twelve __ kisses.
Seven elves came and brought __ kisses.
Eight elves came and brought __ kisses.

Some of the elves can't remember how many kisses they need to bring to the party this year.

Algebra

So Many Children
December

Grade Level: Primary

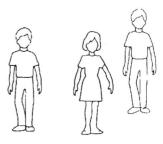

TASK ANALYSIS: 4 – Identifies, uses, and creates true/false number sentences
5 – Identifies, uses, and creates open number sentences
6 – Identifies, uses and creates missing one-digit addends and subtrahends in number sentences
7 – Identifies, uses and creates equations to represent addition and subtraction word problems to show the missing information

MATERIALS: 6-foot long piece of butcher paper, symbols, butcher paper wall chart, marker

ORGANIZATION: Whole group activity
K-3

PROCEDURE:
- Place 6-foot long piece of butcher paper on the floor.
- Children lay the symbols on the paper strip.

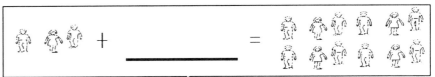

- Read "4 children + ___ children = 12 children."
- Children build (decide how many children need to stand in) missing part of equation.
- Repeat procedure, keeping sum constant but changing the addends.
- Children may record their findings on large butcher paper wall chart.
- Question if we keep "4" here and change "8" to "9", then is the equation true or false?
- Literature: Read: There Was an Old Woman (K-1)
　　　　　　　　　Cheaper By the Dozen (2-3)
　　　　　　　　　The Relatives Came (K-3)

31

Algebra

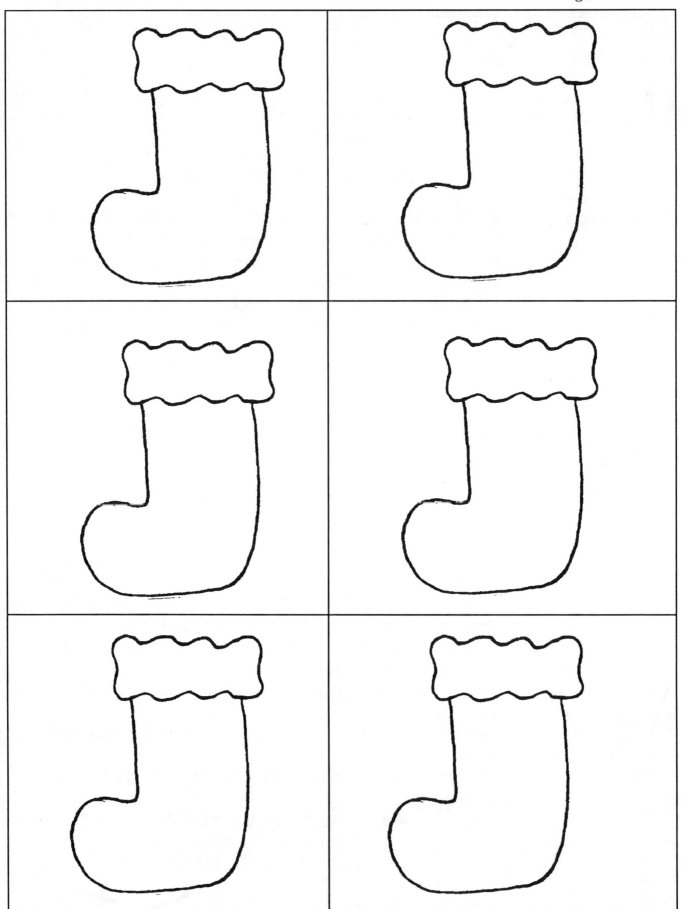

Algebra

What Will You Find in Your Stocking?

December

Grade Level: Primary

TASK ANALYSIS: 1 – Identifies, uses, and creates sets of variables and/or numbers used to replace unknowns
5 – Identifies, uses, and creates open number sentences
7 – Identifies, uses, and creates equations to represent addition and subtraction word problems to show the missing information

MATERIALS: Butcher paper graph 5 to 6 feet long shaped like a Christmas stocking, headings: TOYS, BOOKS, FOOD, CLOTHES, markers, glue, xeroxed stockings (one sheet per child), scissors, pencils

ORGANIZATION: Whole class activity
Cooperative groups for second and third grade
K–3

PROCEDURE: – Cut out 5-6 foot long stocking on butcher paper.

Algebra

Title: What Will You Find in Your Stocking?
Make columns with headings: TOYS, BOOKS, FOOD, CLOTHES.
Discuss with children what they want to find and what they think they may find in their Christmas stockings.
Give children a xeroxed sheet of stockings.
Children cut out the stockings they need for the graph.
Children glue stockings in appropriate columns.
Discuss and interpret data.
Use algebraic symbols to create number sentences.
Cooperative groups write stories using the data from the graph to create word problems.
Children then move round-robin from group to group finding solution sets for each cooperative group's story problem.
As whole group share story problems and solutions.

Algebra

Bears Love Honey
January
Grade Level: Primary

TASK ANALYSIS:
1. Identifies, uses and creates sets of variables and/or numbers used to replace unknowns
2. Identifies, uses, and creates + and − symbols to show relationship between numbers
3. Identifies, uses and creates <, >, and = symbols to show relationship between numbers
4. Identifies, uses, and creates true/false number sentences
5. Identifies, uses, and creates open number sentences
6. Identifies, uses and creates missing one-digit addends and subtrahends in number sentences

MATERIALS: Butcher paper, marker, bear pattern (1 per child), honey, dipper, crackers (round or square)

ORGANIZATION: Whole class activity for presentation
Tasting and graphing at a station
K–3

PROCEDURE:
- Prepare a chart with the heading "Bears Like Honey. Do You?" with Yes/No columns (see next page for graph).
- Teacher introduces graph.
- Teacher models choosing cracker, dipping honey on the cracker, choosing bear and placing on graph.
- Children go through the honey tasting station.
- They drip honey on a cracker of their choice.
- They eat the cracker with honey and answer the question: "Do you like honey?"
- Children attach a bear pattern to the chart under the appropriate heading (yes or no).

Algebra

As a whole group activity ask these questions:
"There are:
___ bears more than ___ bears."
___ bears less than ___ bears."

(number) bears in the yes column."
(number) bears in the no column."
(number) more yes bears than no bears."
(number) less no bears than yes bears."

Children brainstorm for ways to make their number sentences true/false. Discuss and interpret solution sets.

Algebra

37

Algebra

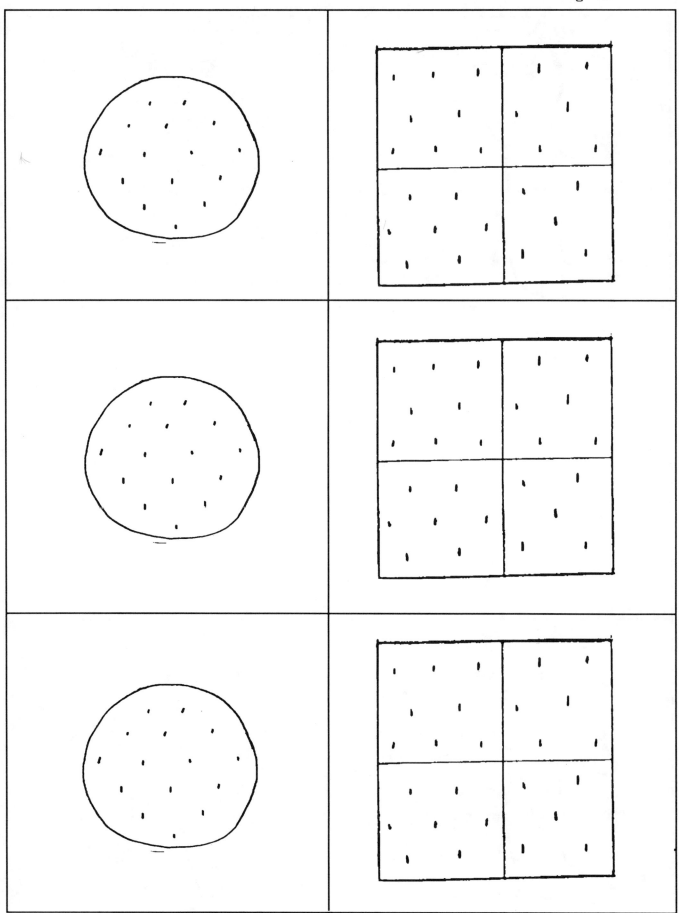

Algebra

Crackered-up
January
Grade Level: Primary

TASK ANALYSIS: 1 – Identifies, uses and creates sets of variables and/or numbers used to replace unknowns
4 – Identifies, uses, and creates true/false number sentences
5 – Identifies, uses, and creates open number sentences
6 – Identifies, uses and creates missing one-digit addends and subtrahends in number sentences

MATERIALS: Two shapes of crackers (i.e. round and square), jelly, butter, two spreaders, butcher paper, markers, scissors, xeroxed crackers (one of each shape per child), glue

ORGANIZATION: Whole class activity for presentation
Graphing done independently (or teacher may assist Kdg.-1)
Whole class activity for problem solving
K-3

PROCEDURE:
- Children choose favorite cracker.
- Children spread their choice of toppings provided.
- Children eat the cracker and then answer the question – What shape was your cracker?
- Children record answer by cutting out the xerox shape of the cracker chosen. (They may or may not choose to color the pattern)
- Mount the cracker symbol on the butcher graph.
- As a whole group activity, brainstorm and share number sentences about the graph created:
    Which column has more crackers?
    Which column has less crackers?

Algebra

How many more square crackers are there than round crackers?
　How many less round crackers are there than square crackers?
　How many square crackers?
　How many round crackers?
Discuss and interpret data.

Extension: The children may graph the data about the toppings they used on the crackers.

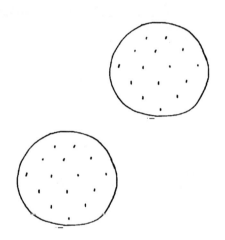
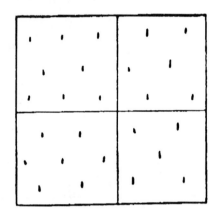

Algebra

Rebus Bear Walk
January
Grade Level: Primary

TASK ANALYSIS: 1 – Identifies, uses, and creates sets of variables and/or numbers used to replace unknowns
2 – Identifies, uses, and creates + and – symbols to show relationship between numbers
3 – Identifies, uses, and creates <, >, and = symbols to show relationship between numbers
5 – Identifies, uses, and creates open number sentences
6 – Identifies, uses and creates missing one-digit addends and subtrahends in number sentences
7 – Identifies, uses and creates equations to represent addition and subtraction word problems to show the missing information

MATERIALS: Butcher paper, markers, drawing paper, pencils

ORGANIZATION: Whole class activity for story reading
Whole class activity for K-1 problem solving
Cooperative groups for 2-3
K-3

PROCEDURE: – Teacher places butcher paper on wall or easel.
– Have markers ready to do drawings of rebus story. (If you are uncomfortable drawing – do drawings before presentation.)
– Teacher begins reading the story to the class.
– Children may "read" along as they see number pattern emerge.
– Children use picture chart the teacher has drawn to create number sentences.
– Children brainstorm for what is missing – what must be added so the bears have what they want?
– Children create equations to show the bears adventure.

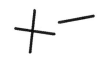

Algebra

Children may wish to create the number sentences and equations using classmates to represent bears, trees, etc.
Then transfer solutions and data to their own picture number problem.
Next transfer to number solution sets.

One day two little (bears) stomped through the autumn leaves as they listened to the birds singing in the (woods) . They were lonely little (bears) .
They wanted children to come to the (woods) so they would each have a playmate.
How many (children) need to go into the (woods) ?

The second day three little (bears) went into the (woods) . They liked the (woods) but they wanted to find enough (beehives) so they could each pull honey from their very own (hives) .
How many (hives) will the little (bears) need to find in the (woods) ?

The third day four little (bears) tiptoed into the (woods) . They listened to the birds singing, to the bees buzzing, but the four (bears) wanted a cool (puddle) in which to splash.
How many (puddles) must the (bears) find?

Continue the story so that the bears go to the woods for five days, but use the story for short periods of time throughout the day or for a period of five days.

Algebra

"100" Bags
February
Grade Level: Primary

TASK ANALYSIS: 1 – Identifies, uses, and creates sets of variables and/or numbers used to replace unknowns
3 – Identifies, uses, and creates <, >, and = symbols to show relationship between numbers
4 – Identifies, uses, and creates true/false number sentences
6 – Identifies, uses, and creates missiong one-digit addends and subtrahends in number sentences
7 – Identifies, uses, and creates equations to represent addition and subtraction word problems to show the missing information

MATERIALS: Butcher paper, black marker, glue

ORGANIZATION: Whole class activity
K-3

PROCEDURE:
- Brainstorm as a class on the 94th or 95th day of school: What objects could we bring 100 of to school?
- Children will spend the next few days gathering their baggie of "100".
- Some ideas our children have discovered are:

elbow macaroni	pennies
chocolate chips	pinto beans
lima beans	popcorn kernels
toothpicks	raisins
pretzel sticks	tiny shells
peanuts	fruit loops
cheerios	crackers

- Make a graph with four columns (no headings).
- Children bring their baggie with 100 items to school on the 100th day and place it on the graph.
- Children decide the catagories as bags are added to the graph.

Algebra

Children may move and re-move baggies until the majority agrees on the reasoning for the placement.
Make headings on the graph.
An example:

Pennies	Toothpicks	Ready to Eat	Must Cook

Ask children to volunteer number sentences about the graph:
"There are more children who brought pennies than any other item."
"There are less bags of toothpicks than ready to eat food."
"There are the same number of things to eat as there are things you can't eat."
"There is one more bag of ready to eat than of toothpicks."
The teacher may then ask children: "How may we change our number sentences to make them false (untrue)?
Second and third grade children may make equations using the data from the graph and the number of children in the room.
Each child or cooperative group may record it pictorially and numerically.
As a whole class discuss and compare findings.

My Silly Valentine
February
Grade Level: Primary

TASK ANALYSIS: 1 – Identifies, uses, and creates sets of variables and/or numbers used to replace unknowns
3 – Identifies, uses, and creates <, >, and = symbols to show relationship between numbers
4 – Identifies, uses, and creates true/false number sentences
5 – Identifies, uses, and creates open number sentences

MATERIALS: Butcher paper graph, one xerox heart per child, glue, marker, pencils

ORGANIZATION: Whole class activity
K-3

PROCEDURE: – Make butcher paper graph divided in half with headings:

Would You Like to Get a	
Silly Valentine	Pretty Valentine
♡ ♡ ♡ ♡	♡ ♡ ♡

Algebra

Have one paper heart per child.
Discuss and look at several pretty and silly valentines.
Ask for children to share how either kind may make them feel inside.
Children put their name on one side and the word silly or pretty on the other side of the paper heart.
Children glue it under the correct heading.
Read and interpret:
   Are there less silly or less pretty valentines?
   Are there more silly or more pretty valentines?
   How many silly valentines are there?
   How many pretty valentines are there?
   How many less silly than pretty valentines?
   How many less pretty than silly valentines?
   How many more pretty than silly valentines?
   How many are there all together?
   Why is it true?
   Why is it false?

Algebra

Algebra

How Many Coins Do Your Crayons Weigh?
Recording Sheet

Estimate Actual > < =

Algebra

How Many Coins Do Your Crayons Weigh?

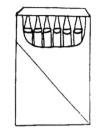

February

Grade Level: Primary

TASK ANALYSIS: 2 – Identifies, uses, and creates + and – symbols to show relationship between numbers

3 – Identifies, uses, and creates <, >, and = symbols to show relationship between numbers

4 – Identifies, uses, and creates true/false number sentences

5 – Identifies, uses, and creates open number sentences

7 – Identifies, uses, and creates equations to represent addition and subtraction word problems to show the missing information

MATERIALS: Balance scales for each cooperative group, box of 8 crayons for each group, bag of coins (pennies, nickles, dimes, quarters) for each group, xerox copy of record sheet (one per group), pencils

ORGANIZATION: Whole group activity for presentation
Cooperative learning groups
K-3

PROCEDURE: – The box of 8 crayons will be the constant for comparing the equations.
– Form cooperative groups.
– Each cooperative group places crayons in balance scale.
– Group estimates the number of coins needed to balance the scale. (This procedure will be done for as many different coins as available.)
– Record estimate.
– Each group adds coins to make scale balance.
– Count and record the actual number of coins needed to balance the scale.

Algebra

Each group gives number sentence using the equations created by comparing the estimate to the actual number of items.
Each group records their own equations.
Two cooperative groups work together to compare and interpret data.

Literature: <u>Abraham Lincoln as a Boy</u>

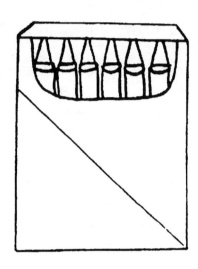

Algebra

Windy Kites
March
Grade Level: Primary

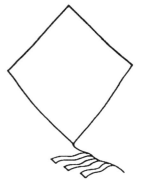

TASK ANALYSIS: 1 – Identifies, uses, and creates sets of variables and/or numbers used to replace unknowns
2 – Identifies, uses, and creates + and – symbols to show relationship between numbers
3 – Identifies, uses, and creates <, >, and = symbols to show relationship between numbers
4 – Identifies, uses, and creates true/false number sentences
5 – Identifies, uses, and creates open number sentences
6 – Identifies, uses, and creates missing one-digit addends and subtrahends in number sentences
7 – Identifies, uses, and creates equations to represent addition and subtraction word problems to show the missing information

MATERIALS: Two task cards, markers, 12" x 18" construction paper, string or yarn, glue, pencils, primary story paper with space for illustrations, crayons

ORGANIZATION: Whole class activity for presentation
Cooperative groups
K-3

PROCEDURE: – Teacher prepares task cards (see examples on following page).
– Teacher has materials ready for children to use.
– Brainstorm as a class what they know about kites and wind.
– Read the task card that has the story starter as a class.
– Children work in cooperative groups to create their number story.
– Children create number sentences based on their story.
– Children write equations based on their story.
– Children draw illustrations on each story page to show the equation and/or number sentence.

Algebra

Note to teacher: We find these stories are usually 4-6 pages in length. The children then do the second task card so as to develop the book cover. Often children will want to then write another "Number Story" so they may create their very own number book to take home.

Some other ideas our children have used are: leprechauns, shamrocks, peanuts, bunnies, eggs, rainbows, and pots of gold.

Extension: Children trade "Number Story Books".
Read story with a partner.
Act out the equation using classmates.

BOOK COVER TASK CARD

1 light blue sheet (12 x 18) for the background
2 gray pieces for clouds
1 black strip for land
5 colored squares for kites
5 pieces of string for kites tails
Glue clouds onto blue background paper
Glue only the bottom edge of land strip
Fold colored squares to make kites
Glue kites onto background (only one flap)
Place a string under each kite and glue it down
Look at the example to help you.

(Note to teachers: I prepare a bookcover ahead to assist the children)

STORY STARTER TASK CARD

Once upon a time there were five kites and one of the kites got stuck in a tree. The little stuck kite thought and thought. Then he got an idea of how he could get out of the tree. So he . . .

Algebra

Peanuttiest

March

Grade Level: Primary

TASK ANALYSIS: 1 – Identifies, uses, and creates sets of variables and/or numbers used to replace unknowns
3 – Identifies, uses, and creates <, >, and = symbols to show relationship between numbers
4 – Identifies, uses, and creates true/false number sentences
5 – Identifies, uses, and creates open number sentences

MATERIALS: Attribute hoops or yarn circles, tied baggies of shelled peanuts, tied baggies of chocolate-covered peanuts, tied baggies of chocolate-covered raisins (several of each so that there is one per child), tagboard symbols of <, >, =, equations mat

ORGANIZATION: Whole class activity
K-3

PROCEDURE:
- Put the equation mat on the floor.
- Place attribute hoop or yarn on equation mat.

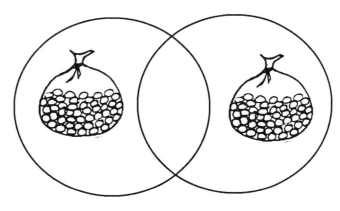

- Teacher places bag of peanuts in a hoop.
- Then she places bag of chocolate-covered raisins in the other hoop.
- Then one at a time children place their baggie in the hoop they think is appropriate.

Child gives sentence telling why they put their baggie in that hoop.
No corrections are made at this point because children need to discover the sets and subsets.

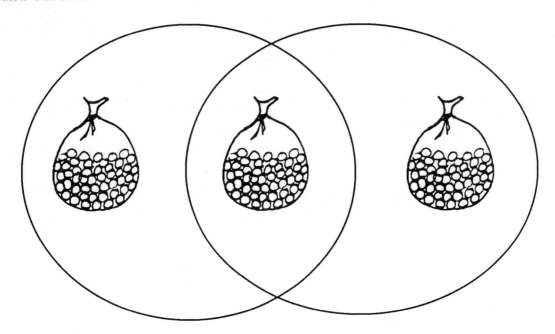

After all the baggies are placed, then the teacher does the following questioning:
  "Does anyone see any bags they want to place differently?" (Usually children see needed changes)
   "What's the same about all the baggies in this set?" (chocolate-covered raisins)
   "What is different about the baggies in this set?"
   "Where might the different baggies belong?"
(Allow children plenty of time to move and re-move from set to set.)
Then do the same questioning for the other set (shelled peanuts).
Now you are ready to discuss the subset (created by the chocolate-covered peanuts).
Do similar questioning for the subset.
Children, holding baggies and symbols, create a live equation.
Then children transfer the baggies and symbols to the equation mat.
Have children create true/false equations.

Literature: George Washington Carver

Algebra

Tater Time
March
Grade Level: Primary

TASK ANALYSIS: 1 – Identifies, uses, and creates sets of variables and/or numbers used to replace unknowns
2 – Identifies, uses and creates + and – symbols to show relationship between numbers
3 – Identifies, uses and creates <, >, and = symbols to show relationship between numbers
4 – Identifies, uses, and creates true/false number sentences

MATERIALS: Balance scales, potatoes, record sheet, pencils and markers, beans, tiles, wooden cubes

ORGANIZATION: Whole group activity for presentation
Cooperative groups
K–3

PROCEDURE: – Teacher may choose to model weighing a potato with chosen unit of measure.
– Groups choose the one arbitrary measure they will use. This must remain constant.
– Also each cooperative group marks their potato so as to distinguish it from all other potatoes.
– Cooperative group places potato in balance scale.
– Group then estimates number of tub items needed to balance the scale.
– Record estimate.
– Proceed to balance scale.
– Group counts and records <u>actual</u> number of items needed to balance the scale.

Algebra

Group compares estimate and actual count to create an equation.
        We estimated < actually needed.
        We estimated > actually needed.
        We estimated = as actually needed.
Groups share data with whole class.

Extension: Have two cooperative groups keep same unit of measure but exchange potatoes.
Repeat procedure: estimate; record; balance-count actual number; record; create equation from data.
The groups then compare and interpret data.

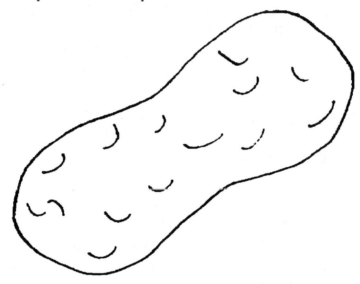

Algebra

Tater Time
Recording Sheet

Estimate   Actual   >   <   =

Algebra

Rain, Rain Go Away
April
Grade Level: Primary

TASK ANALYSIS: 1 – Identifies, uses, and creates sets of variables and/or numbers used to replace unknowns
2 – Identifies, uses, and creates + and − symbols to show relationship between numbers
3 – Identifies, uses, and creates <, >, and = symbols to show relationship between numbers
5 – Identifies, uses, and creates open number sentences

MATERIALS: Umbrellas and slickers, yarn for circles, equation mat and tagboard symbols: +, −, <, >, =

ORGANIZATION: Whole class activity
K-3

PROCEDURE:
- Ask children the day before to come to school dressed for rain with umbrellas, slickers or both.
- Create large yarn circles on floor.
- Teacher places a child with an umbrella in one circle.
- Place a child with a slicker and umbrella in another circle.
- Then the other children go stand in the circle they think is appropriate.
- Teacher then asks each child to volunteer the reason for their choice.
- Ask for a volunteer to place the children as he/she thinks they should be placed.
- Have the child explain his/her reason.
- Note to teacher:
- Continue with this process until children are standing in the following venn diagram.

Algebra

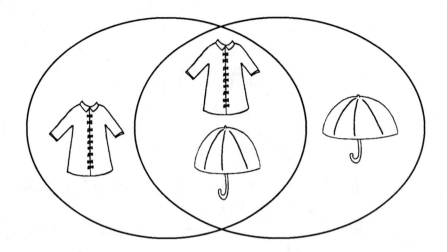

Next have children create number sentences standing on the equation mat.

3 umbrellas + 2 slickers = 2 slickers + 3 umbrellas

Children brainstorm as to why this is true.
Children then continue to create equations using +, −, <, >, and =.
Discuss and interpret data.

Algebra

How Heavy is Your Egg?
April
Grade Level: Primary

TASK ANALYSIS: 2 – Identifies, uses, and creates + and – symbols to show relationship between numbers
3 – Identifies, uses, and creates <, >, and = symbols to show relationship between numbers
4 – Identifies, uses, and creates true/false number sentences
5 – Identifies, uses, and creates open number sentences
7 – Identifies, uses and creates equations to represent addition and subtraction word problems to show the missing information

MATERIALS: Balance scales for each cooperative group, tub of keys, tub of tiles or tub of junk crayons for each group, one hard-boiled egg per child, xerox copy of record sheet (one per group), pencils

ORGANIZATION: Whole group activity for presentation
Cooperative learning groups
K–3

PROCEDURE: – Class decides which arbitrary units of measure will be used. This must be the constant for comparing the equations.
– Model weighing an egg in the balance scales with the chosen unit of measure if your class has not had this experience.
– Form cooperative groups.
– Each child in the cooperative group does the following while other group members observe and discuss variables:
    Place egg in balance scale.
    Estimate the number of tub items needed to balance the scale.
    Record estimate.
    Child adds items to make scale balance.

Algebra

After each child in the cooperative group has recorded their data:
Each group gives number sentences using the equations created by comparing the estimate to the actual number of items.
Each group records their own equation.
Two cooperative groups work together to compare and interpret data.

Second and third grade children may extend this by creating a wall chart showing the variables of all cooperative groups the equations and the solution sets.

Algebra

How Heavy Is Your Egg?
Recording Sheet

Estimate	Actual	> < =

Algebra

Seedy Business
May
Grade Level: Primary

TASK ANALYSIS: 2 – Identifies, uses, and creates + and – symbols to show relationship between numbers
3 – Identifies, uses, and creates <, >, and = symbols to show relationship between numbers
4 – Identifies, uses, and creates true/false number sentences
5 – Identifies, uses, and creates open number sentences
7 – Identifies, uses and creates equations to represent addition and subtraction word problems to show the missing information

MATERIALS: Balance scales for each cooperative group, tub of unifix cubes, tub of brads or tub of paper clips for each cooperative group, one cup of sunflower seeds per cooperative group, xerox copy of record sheet (one per group), pencils

ORGANIZATION: Whole group activity for presentation
Cooperative learning groups K–3

PROCEDURE:
- Class decides which arbitrary unit of measure will be used by each cooperative group. This must be the constant for comparing the equations.
- Form cooperative groups.
- Each cooperative group places sunflower seeds in balance scale.
- Group estimates the number of tub items needed to balance the scales.
- Record estimate.
- Each group adds items to make scale balance.
- Count and record the actual number of items needed to balance.
- Each group gives number sentence using the equation created by comparing the estimate to the actual number of items.
- Each group records their own equation.
- Two cooperative groups work together to compare the solution sets.

Algebra

Seedy Business?
Recording Sheet

Estimate	Actual	> < =

Algebra

What's True for You?
May
Grade Level: Primary

TASK ANALYSIS: 3 – Identifies, uses, and creates <, >, and = symbols to show relationship between numbers
5 – Identifies, uses, and creates open number sentences

MATERIALS: Butcher paper, markers, venn diagram, headings, adhesive dots, tally sheet

ORGANIZATION: Task card done individually
Whole class activity
Cooperative groups
K–3

PROCEDURE:
- Make task card: Do you like puppies or kitties or both?
- Place on door so when children come to class, they can read the task and place their dot on the venn diagram on the wall.

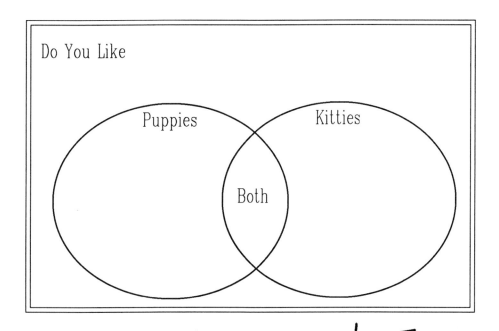

Algebra

Children create number sentences using <, >, =.
Teacher may use the following questions to facilitate children's responses:
"How many liked puppies?"
"How many liked Kitties?"
"Do more like puppies than kitties?"
"Do less like puppies than Kitties?"
"Do more people like both puppies and kitties instead of just puppies?"
Children may mark tally sheet as they share findings.

Puppies	Kittens	Both
IIII	IIIII	IIII

Discuss and interpret data.

Note: Second and third graders may wish to work independently or in cooperative groups to do the above procedure. Then they share their solutions with the whole class.

We have had our K-1 children bring stuffed puppies and kitties to create their floor venn diagram. We integrate the activity with our Social Studies and Science units on the care of pets, their importance in our lives, animal life cycles, birth/death. The children transfer the floor diagram data to wall graphs, bar graphs and create little books of "Pet Story Problems".

Algebra

What Did You Like Best About Language Arts?

My Journal

June

Grade Level: Primary

TASK ANALYSIS: 1 – Identifies, uses, and creates sets of variables and/or numbers used to replace unknowns

2 – Identifies, uses, and creates + and – symbols to show relationship between numbers

3 – Identifies, uses, and creates <, >, and = symbols to show relationship between numbers

6 – Identifies, uses, and creates missing one-digit addends and subtrahends in number sentences

7 – Identifies, uses, and creates equations to represent addition and subtraction word problems to show the missing information

MATERIALS: Adhesive dots, task card on classroom door, 12" x 18" paper, markers/pencils, venn diagram on wall by task card, butcher paper

ORGANIZATION: Whole class activity
Cooperative groups to create equations and word problems
K-3

PROCEDURE:
- Children read task card on door as they enter the classroom:
- What was your favorite activity in Language Arts this year?
- Children place dot to show what they enjoyed.
- K-1 as a whole group and 2-3 as cooperative groups discuss and interpret data.
- Have bar graph hanging on wall.
- Shade in class bar graph to show solutions.

Algebra

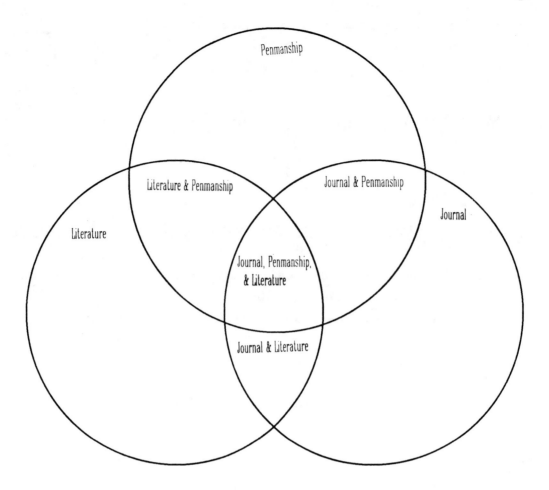

Using data from bar graph children create addition and subtraction number sentences.
Then children create numerical equations based on word problem data.
As whole class activity share story problems and solution sets.

Algebra

What Do You Like Best About Language Arts?
Recording Sheet

Journal	Journal & Literature	Journal & Penmanship	Literature	Literature & Penmanship	Journal, Penmanship, & Literature	Penmanship

Algebra

I Scream, You Scream
Recording Sheet

Estimate	Actual	> < =

Algebra

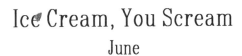

Ice Cream, You Scream
June
Grade Level: Primary

TASK ANALYSIS: 2 – Identifies, uses, and creates + and – symbols to show relationship between numbers
3 – Identifies, uses, and creates <, >, and = symbols to show relationship between numbers
4 – Identifies, uses, and creates true/false number sentences
6 – Identifies, uses and creates missing one-digit addends and subtrahends in number sentences

MATERIALS: Balance scales for each cooperative group, tub of unifix cubes, tub of tiles, or tub of pattern blocks for each cooperative group, one flat bottom cone per child, ice cream, several ice cream scoops, xerox copy of record sheet (one per group), pencils

ORGANIZATION: Whole group activity for presentation
Cooperative learning groups for dipping and weighing activity
K-3

PROCEDURE:
- Class decides which arbitrary unit of measure will be used by each cooperative group.
- This must be the constant for comparing the equations.
- Model weighing an ice cream cone (one dip) in the balance scale with the chosen unit of measure.
- Form cooperative groups.
- Each cooperative group places ice cream in cones.
- Then place one cone in balance scale.
- Children estimate the number of tub items needed to balance the scales.
- Record estimate.

Algebra

Each group adds the tub items to make scale balance.
Count and record the actual number of items needed to balance the scale.
Each group gives number sentence using the equation created by comparing the estimate to the actual number of items.
Each group records their own equation.
Two cooperative groups work together to compare the variables and solution sets.
Groups interpret data and report to whole class.

Algebra

| 8 | Recognizes and identifies variables and equations |

Equalities in Advertising
Grade Level: Middle

MATERIALS: Magazines, scissors, glue, construction paper

ORGANIZATION: Individually or in teams of two

PROCEDURE: This lesson will give students a very basic understanding of the purpose of an equation and the meaning of a variable.

Begin by explaining to the whole class that part of the word "equation" means "equal." In math, when two sides of a number sentence are separated by an equal symbol, it means that there is an equal amount on either side of the symbol.

Write the number sentence 3 x 5 = 15 on the board and ask a student to read it aloud for the class. The student will either use the word "equals" or "is" to describe the equal symbol. Tell students that in most advertisements, there are equations which are written in words rather than in numbers. Students will be searching magazines, cutting out word equations and then writing these in simplified forms. Several examples are shown below.

"The Choice is clear!"
Choice = clear
Choice is a variable because several choices might be made and still be clear

-"As Individual As You Are"
Individual = You
Is there a variable? This would be open to debate.

Once they have done this portion of the lesson, have students go through their equations and identify possible variables (items which could be changed within the advertisement). They will find that most advertisements also have variables.

This integration of a math concept with language will help students remember the purpose of equations and variables.

Hands On, Inc
2121 Rebild Drive
Solvang, CA 93463

Algebra

| 8 | Recognizes and identifies variables and equations |

Cheeseburger, Cheeseburger
Grade Level: Middle

MATERIALS: No special materials are necessary

ORGANIZATION: Cooperative groups of four

PROCEDURE: Explain to your students that algebra is used to represent situations in real life that can be solved mathematically. Two key terms in algebra are "variable" and "equation." This lesson will help show students the way in which variables function within equations.

Variables are used to represent quantities which change in different situations. An equation represents a set of variables which have a specific relationship. For example, Length multiplied by Width will result in the measurement of the area of a rectangle. This can be represented by the algebraic equation $a = l \times w$. No matter how much l and w change, they will always equal a (area) when they are multiplied.

In this activity students are to imagine that a new hamburger chain has just opened in town. They supposedly serve the "best food ever." Ask the class for suggestions of what the restaurant serves, and put the menu on the board. A sample might be:

```
plain burger(p) = $2.00        tomatoes(t) = .25       salad(s) = .75
extra burger patty(e) = $1.00, onions(o) = .25         soda(a) = .75
extra cheese(c) = .50          french fries(f) = .75   dessert pies(d) = $1.50
lettuce(l) = .25               onion rings(r) = .75
```

Tell students they must order a burger with at least three items on it, one extra item, and a drink. The pie is optional. They must express the order as an equation so their classmates will be able to total their bill. A sample equation might be: $b + c + t + r + s = ?$ Have students circulate their orders around the cooperative groups for each member to solve.

Hands On, Inc
2121 Rebild Drive
Solvang, CA 93463

Algebra

| 8 | Recognizes and identifies variables and equations |

Playing a Part
Grade Level: Middle

MATERIALS: Magazines or crayons, paper, glue, and stapler

ORGANIZATION: Individually

PROCEDURE: This is a basic lesson on recognizing a variable and understanding how it fits into an equation. Students must understand that a variable is an element in an equation which can change in certain situations without affecting the final results/relationships of the equation.

Ask your students for examples which they might see in everyday life where there is a great variety of choice but the end result is the same no matter what choice is made. After some discussion (and confusion) ask them if there is a great variety among themselves, yet they are all humans. No matter how the various parts of people vary, the end result is always a person.

Explain that they will be creating a "paper person" from magazine cutouts. Their job will be to cut out all the basic parts of a person, from separate pictures, and glue them down so as to form a sort of "mosaic" man or woman. Next, they should select one part of the body to be a variable and then find at least five more pictures of that variable in the collection of magazines. These should be stapled to the mosiac-person so that they can be flipped yet the equation stays the same in spite of the "value" of the variable.

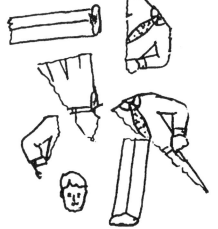

For example, a picture could be constructed from the legs of a fashion model, the torso of an athlete, the arms from a news picture and the head from any five pictures. The head would be the VARIABLE. It is also possible to do this activity by drawing the different parts instead of using the cut-and-paste method.

Hands On, Inc
2121 Rebild Drive
Solvang, CA 93463

Algebra

| 8 | Recognizes and identifies variables and equations |

Red and Yellow Make...Variables!
Grade Level: Middle

MATERIALS: Tempera paint, brushes, paper plates, wheels as shown below

ORGANIZATION: Whole class activity to be done by each student

PROCEDURE: This activity uses an art activity to help students identify sets of variables.

Give each student a paper plate for mixing colors, a paint brush, the color wheels as shown below, and blue, yellow, red, and white paint. Explain to the students that their task is to identify the set of colors which will create the colors in the color wheels. For example, red and white create pink, or blue and yellow create green.

On the board write: A = red, B = yellow, C = blue, D = white. Let the students explore with paint mixtures to find the right combination (set) of colors. Once the area is colored appropriately, have the students write the correct equation for each color on that section of the wheel.

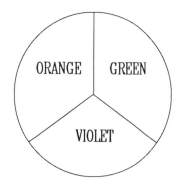
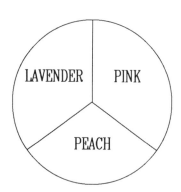

Algebra

| 9 | Identifies basic level symbols: <, >, = |

Wheeling and Dealing
Grade Level: Middle

MATERIALS: Deck of cards, dice, or dominoes

ORGANIZATION: Teams of two students

PROCEDURE: One of the skills needed in algebra is the ability to recognize relationships between numbers and the quantities they represent. In this activity students will practice their skills using various symbols to make number sentences true.

Model for the class by drawing three cards from the deck and writing the values on the board (ace = 1, face cards = 10). Now place a multiplication sign between the first two numbers and ask the class to finish (make true) the number sentence by adding the proper sign (=, <, or >).

Tell your students that after they have done one number sentence for multiplication, they should do one for each of the other operations. Some sample sentences are shown below.

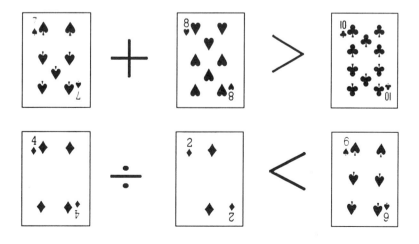

Algebra

| 9 | Identifies basic level symbols: <, >, = |

Know It All's
Grade Level: Middle

MATERIALS: Three dice for each pair of students, index cards with =, <, > symbols, large cards for teacher explanation

ORGANIZATION: Pairs of students

PROCEDURE: In this activity students will be using various symbols to make random number sentences true. They will also be practicing the skill of solving for an unknown number.

Begin by modeling this procedure with your class. Roll three dice, but do not show the numbers to the class. Next write the equation A + B ☐ C on the board. Their job will be to first, guess the value of each die. To do this, call on a student to guess a number; the teacher should respond by holding up the appropriate card (<, >, =) thus narrowing the choices. Continue until students have identified the three numbers.

Second, replace the variables (A, B, and C) with the numbers on the dice and have students select the symbol which will make the number sentence true.

Once they are familiar with this procedure, divide the class into pairs and give each team three dice. Designate one student to be the "know-it-all" and the other, the guesser. Give a set of symbol cards (<, >, =) to each "know it-all" and have students repeat the procedure outlined above.

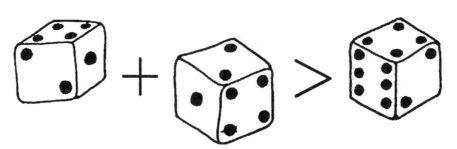

Hands On, Inc
2121 Rebild Drive
Solvang, CA 93463

Algebra

| 9 | Identifies basic level symbols: <, >, = |

Picture This!
Grade Level: Middle

MATERIALS: Catalogs, scissors, glue, paper for books, stapler

ORGANIZATION: Groups of two

PROCEDURE: Explain to students that they are going to make a book that uses symbols to show relationships. Prior to the activity the teacher should cut ditto paper in half to make pages for students' books. The number of pages will be determined by the number of symbols for which the child will create problems.

Demonstrate in front of the class how the child is to proceed. For example, the teacher should cut out three dolls from a catalog, glue them to the page and draw a box for the appropriate symbol (=) with three more dolls glued to the right side of the box. Do not put the symbol in the box. Repeat procedure with a similar item to show a "greater than" (>) problem.

Next, list the symbols on the board for which the students are to create a visual problem on their pages. Tell students not to do them in the same order listed on the board. The last page of the book should contain symbols in the correct size to fit the box in each problem. Make an attractive cover for the book and staple the completed book together.

As a final step, have groups exchange books. Tear out the last page, cut the symbols out and glue them to the appropriate equation.

In having to create the problem and find answers to another problem the student does a double practice.

 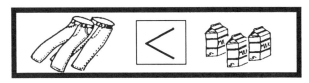

Hands On, Inc
2121 Rebild Drive
Solvang, CA 93463

Algebra

9	Identifies basic level symbols: <, >, =

Calling Your Own Number
Grade Level: Middle

MATERIALS: HANDO cards (see Appendix A) and markers

ORGANIZATION: A whole class activity

PROCEDURE: This is a bingo type game played using HANDO cards. The purpose of the lesson is to have students generate number sentences which use greater than and less than terminology.

Pass out the HANDO cards, one to each student. Have the students print a number (less than 50) in each square in a random fashion. Tell students that this game is a bit different from bingo in that each student will have a chance to be the "caller" during the game. Another twist to this game is that students may "cover" more than one square at a time on their HANDO card.

The teacher should begin with the following example: "Place a marker on the numbers which are greater than 3 x 5 and less than 2 x 9." Guide students to place markers on the numbers 16 and 17 -- not on 15 or 18 since they are EQUAL TO the given equations. A second type of statement might be, "Place a marker on the number with is equal to 4 x 9."

Once students have the concept, ask for a volunteer to give a set of statements. Continue with this procedure by either calling on students who volunteer or by rotating around the classroom in a set pattern having each student give a statement.

The student who fills in the entire card is the winner. You will find that as students play the game, they will begin to use strategies to call numbers which they can use to complete their game cards.

H	A	N	D	O

Hands On, Inc
2121 Rebild Drive
Solvang, CA 93463

Algebra

| 10 | Identifies sets and/or numbers used to replace variables |

Rollin' Around
Grade Level: Middle

MATERIALS: Gameboard (Appendix B), chart of letters and numerical equivalents, dice and markers (beans)

ORGANIZATION: Groups of two, three, or four students

PROCEDURE: This is an exercise to practice substituting numbers for variables.

The teacher should explain the game and have a practice game, then allow the students to play without help. Tell the students to group themselves in 2's, 3's, or 4's to play the game.

Hand out the gameboards, dice, and markers. Each group or player can have a list of the letters with their numerical equivalents, or the list can be posted for all to refer to. The list can be as follows or the equivalents can be changed or shuffled (A = 1, B = 2, C = 3, etc.).

Each group plays on one gameboard. Each player rolls the die and the player with the highest number begins first and play moves to the left. Each player rolls the die (in turn) to move the marker on the board. If players roll a two, they can move to the letter B on the board; if they roll a three, move to C. Play proceeds forward to the next letter on the board. If a second player lands on the same letter, the first player goes back to the start (only one player at a time on each space). The one reaching the end first wins.

As an extension, this game can be reversed by using numbers on the gameboard. Print letters on small slips of paper and place them in a box or bag. Players draw the slips and move to that space on the gameboard.

$A = 1$
$B = 2$
$C = 3$
$D = 4$
$E = 5$
$F = 6$

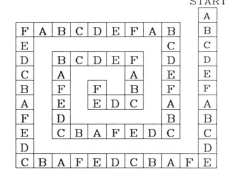

Hands On, Inc
2121 Rebild Drive
Solvang, CA 93463

Algebra

| 10 | Identifies sets and/or numbers used to replace variables |

Cryptograms
Grade level: Middle

MATERIALS: Chart of letters and their numerical equivalents (Appendix C)

ORGANIZATION: Groups of two students

PROCEDURE: The purpose of this lesson is for the students to practice substituting numbers for letters. The teacher should post the chart and demonstrate by writing a code on the board. See example below.

Have the class decode the message and then call on the students to help write a message in code as a class activity. When the students understand how to create codes, have them choose partners. Each partner should make up a code and then exchange it with the partner to decode. The difficulty can be increased by using problems instead of just one number.

As students become proficient with the chart, they can make up new codes and write messages in the code they have created. They must be sure to give the "decoder" a chart of the new code equivalents.

```
   A = 1                    D = 4 + 5
   B = 2                    E = 10 + 11
   C = 3                    F = 12 - 2

3, 1, 2 = C A B          2, 21, 9 = B E D
```

As an extension, you might discuss cryptograms and their use throughout history as secret codes, etc.

Hands On, Inc
2121 Rebild Drive
Solvang, CA 93463

Algebra

| 10 | Identifies sets and/or numbers used to replace variables |

Dollars and Sense
Grade Level: Middle

MATERIALS: Play money (Appendix D – coin page), money equivalent chart (below)

ORGANIZATION: A whole class activity, then teams of two

PROCEDURE: Prepare a large chart or draw the sample chart (below) on the chalkboard. The chart should show pennies, nickels, dimes, quarters, half dollars, and dollars. For older students combinations of five, ten, twenty dollars, or more can be used.

The teacher should begin by writing a monetary amount on the chalkboard or on a slip of paper and give each group of students a supply of coins. Ask students to determine which coins and/or bills would total this given amount of money. The teacher should point out that there will probably be more than one correct answer.

Stipulate that the students are to find as many solution sets as possible, one of which should identify the fewest number of coins and bills necessary. The key is, however, that students must make their representations in the form of an equation. Letter equivalents can be shown as P = 1 cent, N = 5 cents, etc. The equations should be recorded on a chart.

Given a total amount of $.36 cents, some of the sample solutions might be:

$$D+D+D+N+P = 36$$
$$2D+3N+P=36$$
$$Q+D+P=36.$$

Once students understand the concept, divide them into teams of two and have them suggest monetary amounts to one another.

	$.01	$.05	$.10	$.25	$.50	$1.00
Pennies	1	5	10	25	50	100
Nickles		1	2	5	10	20
Dimes			1		5	10
Quarters				1	2	4
Half Dollars					1	2
Dollars						1

Hands On, Inc
2121 Rebild Drive
Solvang, CA 93463

Algebra

| 10 | Identifies sets and/or numbers used to replace variables |

The Puzzling U.S.
Grade Level: Middle

MATERIALS: Puzzles or maps of the United States

ORGANIZATION: Cooperative groups of four

PROCEDURE: Before students work with numbers to solve for unknowns and variables, it is helpful if they can use sentences to go through this procedure. This activity provides a concrete use of language for algebraic expressions.

Divide the class into cooperative groups of four and give each group a map of the United States. Begin by writing the following statement on the board: _____ are states that are bordered by the Pacific Ocean. Ask students if there are several possible answers. Tell them that each of the possible correct answers is a VARIABLE and that all of the correct answers are called the SOLUTION SET. Have the students find the states which make this first statement true (California, Oregon, Washington, Hawaii and Alaska). Ask a student to read the statement on the board providing the correct solution set.

At this point, emphasize the concept that in algebra, there are often numerous answers which can function as variables in a given problem.

Tell students that their task is to create five statements with unknowns (variables), write them down, and then circulate this list among the cooperative groups for solution by class members. When groups solve these statements, they should emphasize the terminology of VARIABLE and SOLUTION SET. Some sample statements might include:

        _____ are states that border Canada.
        _____ are states that touch the Mississippi River.
        _____ are states that border Mexico.

Hands On, Inc
2121 Rebild Drive
Solvang, CA 93463

Algebra

| 11 | Identifies true or false statements or number sentences |

What You See Is What You Get
Grade Level: Middle

MATERIALS: Pictures of how to sign various symbols and numbers from the international signing language (see Appendix E)

ORGANIZATION: A whole class activity and then in teams of two

PROCEDURE: Students are usually fascinated by the sign language used by deaf people. Being able to talk with your hands is something that most students find interesting and motivational. We'll use this motivating aspect in this lesson to help students learn to sign algebraic equations to one another.

Encourage students to start very simply with examples such as 2 + X = 5 but with time they will become more sophisticated with samples such as 2 + 5 is greater than 1 + X therefore X = ?. The response being all numbers less than five.

Begin by teaching the students the signs for the various numbers and symbols. Use basic signs to begin with, but with time you can use signs for parentheses or even brackets.

Let teams of two practice together to create sign equations to present to the class as a whole. Then let the class respond to the equations in sign language.

Algebra

| 11 | Identifies true or false statements or number sentences |

State the Whole Truth
Grade Level: Middle

MATERIALS: Large map of the U. S. (or individual student maps)

ORGANIZATION: Whole class activity, then to cooperative groups of four

PROCEDURE: One of the skills in algebra is the use of letters, called variables, to represent components in an equation. In many instances a variable has more than one value. This activity will give students practice in identifying sets of variables as accurate (true) or inaccurate (false).

Begin by asking your class for the name of a state which borders on the Pacific Ocean (California, Oregon, or Washington -- Hawaii and Alaska as well). In this context, all five of these states are equal, they all border the Pacific Ocean and therefore any of these three answers will make a TRUE statement. By the same token, any other answer will make this statement false.

Ask students to change the original statement to make it false. Some samples might be:

Original: Alaska is a state which borders on the Pacific Ocean. TRUE
Changed: Alaska is a state within the continental United States which borders on the Pacific Ocean. FALSE

Original: California is a state which borders on the Pacific Ocean. TRUE
Changed: California is a state which is surrounded by the Pacific Ocean. FALSE

These are simple examples but in this activity, students are going to create various statements which are true or false and will then ask group members to change them to the opposite meaning. Give students time to create three to five statements and then circulate them through the group.

Algebra

| 11 | Identifies true or false statements or number sentences |

Stand Up for Sentencing
Grade Level: Middle

MATERIALS: Large cards (for each student) with a different number or symbol written on each side

ORGANIZATION: Whole class activity

PROCEDURE: In this activity, the whole class will work together to create number sentences which are true. They will do this by holding up various cards in the front of the room.

Begin by giving each student a large card. Have students write a number (less than 25) on one side of the card and on the reverse side, have them write another number or draw a symbol, such as <, >, =, +, −, x, or ÷.

Call on a student to come to the front of the room with a number card and hold it for all to see. Ask a second student to come forward and do the same. Now comes the challenge. Ask a student who has an operation symbol on a card to come forward and stand between the two numbers. The operation must have a solution of less than 25. Once these three students are in place, ask two more students to come forward and complete the number sentence to make it true. At this point you will have five students in the front of the room and they will be displaying a TRUE number sentence.

At this point, ask one of the symbol students to turn their card over and ask the class if the new number sentence is true or false. If false, ask for a volunteer to come forward to provide a symbol or number which make the sentence true once again. Continue this procedure, pointing out the concept of True and False number sentences.

A sample set of student responses might appear as follows:

Algebra

| 11 | Identifies true or false statements or number sentences |

The Domino Effect
Grade Level: Middle

MATERIALS: Sets of dominoes for each group (students can make their own from cardboard)

ORGANIZATION: Cooperative groups of two or four

PROCEDURE: Students should have a knowledge of basic addition, subtraction, multiplication, and division facts through 6 to be successful at this activity. The purpose is to play a game of dominoes, but rather than matching numbers, they must make number sentences true.

Divide the class into groups and give each group a set of dominoes. Show students that on each side of a domino is a number (of dots). Students must imagine that there is an operation symbol between these two numbers. In order for a student to make a play, they must make a <u>true</u> number sentence from one of the previously played blocks. A sample set of moves is shown below.

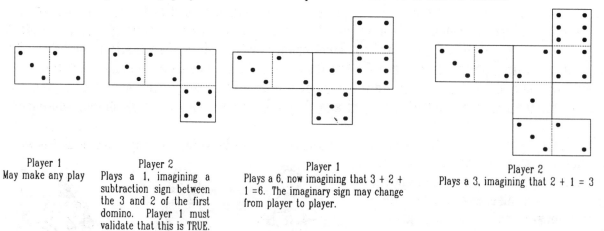

Player 1
May make any play

Player 2
Plays a 1, imagining a subtraction sign between the 3 and 2 of the first domino. Player 1 must validate that this is TRUE.

Player 1
Plays a 6, now imagining that 3 + 2 + 1 = 6. The imaginary sign may change from player to player.

Player 2
Plays a 3, imagining that 2 + 1 = 3

Spread the dominoes, face up on the desk. Students may select from the entire display. The game continues until one player says he cannot play or until all dominoes have been used. We suggest that blank dominoes be removed from the set for this game.

Hands On, Inc
2121 Rebild Drive
Solvang, CA 93463

Algebra

| 12 | Identifies complex symbols: parentheses, brackets, intersections, etc. |

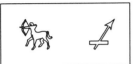

What's Your Algebraic Sign
Grade Level : Middle

MATERIALS: A horoscope prediction from a newpaper -- copy for each student or team

ORGANIZATION: Groups of two students or individually

PROCEDURE: In this activity, students will create a horoscope of their own, but instead of using the signs of the zodiac, they will use the signs of algebra. For example, Sagittarius might be replaced by the "greater than" sign.

The challenge for students is to rewrite the horoscopes providing a prediction which coincides with the meaning of the algebraic sign. There is obviously a lot of room for creativity, but you will also need to lead your class to discover a process to write these horoscopes.

Some samples which you might provide:

If you were born under the sign of GREATER THAN, the number on your open side is bigger than the number on your pointed side. Beware of small numbers trying to trick you into thinking that they should move to the other side.

If your sign is EQUAL TO it means that you have an even temperament and are even on both sides. Don't allow yourself to be fooled by large numbers on one side as long as you can add the numbers on the other side to equal this amount.

You can select from a variety of symbols. If students really get creative, they can do illustrations similar to those found on the actual horoscope. When they have finished, have them share their creations with the whole class.

Hands On, Inc
2121 Rebild Drive
Solvang, CA 93463

Algebra

| 12 | Identifies complex symbols: parentheses, brackets, intersections, etc. |

Serve Up a Symbol
Grade Level: Middle

MATERIALS: Sets of question and symbol cards (Appendix F)

ORGANIZATION: Teams of two students

PROCEDURE: This activity is a game to provide practice in recognition of algebraic symbols.

Begin by reviewing the following symbols with the students:

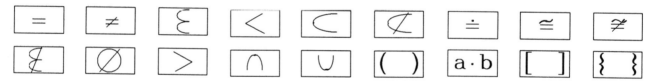

Divide the class into teams of two and give each pair a stack of question cards and symbol cards (see Appendix F). Tell students that they are going to play a game in which the person who selects the correct symbol card will get a point, and five points wins the game. Player 1 selects a question card and displays it for player 2. The second player must choose a symbol that will complete the mathematical statement.

The player holding the question determines if a correct symbol (there may be several correct answers) has been selected. If the response is correct, player 2 receives a point and gets to keep his turn as a guesser. If player 2 is incorrect, player 1 becomes the guesser. This is similar to "winning the serve" in volleyball, in that you cannot score a point unless you are the guesser.

Questions answered correctly are placed in one stack while questions missed are placed in another stack. The winner is the first player to get five points.

As an extension, have the students make a set of questions for each symbol and exchange with another group of players.

Algebra

| 12 | Identifies complex symbols: parentheses, brackets, intersections, etc. |

Sentence Savvy
Grade Level: Middle

MATERIALS: Index cards and markers

ORGANIZATION: Groups of two

PROCEDURE: Mathematicians use parentheses, brackets, and braces to group similar or related information. This helps them know what information goes together and should be dealt with as a single unit. In this activity students will use the parts of speech to write equations and group related information.

Begin by giving twelve index cards to each pair of students. Have the teams write words on each card so that they have three nouns, three verbs, three adjectives and three adverbs (for simplicity have them write words ending in "ly"). They should label each card with a letter/variable so that they can easily be recognized if the cards are shuffled (nouns = N, adverbs = Adv, etc.).

Have students use a noun and verb card to form a simple sentence ("cat runs") and tell them they needn't worry about articles. The equation for this sentence would be N + V = S (sentence). Next, have them add an adjective and adverb ("fat cat runs quickly") and write an equation for this sentence, i.e. Adj + N + V + Adv = S. Ask if the order of these words can be rearranged and guide them to the discovery that Adj + N fit together and V + Adv fit together. Explain that in math, sometimes certain combinations of numbers must also "fit together" and to do this parentheses are used. The sample sentence could be written as [(Adj + N) + (V + Adv)] = S and since all four parts of speech work together, the entire sentence might have brackets.

Let the students experiment with their other cards, creating sentences with several adjectives or adverbs. For each sentence created, they should write the sentence as an equation. You can extend this into a discussion of compound and complex sentences to more clearly explain the use of brackets and braces.

Hands On, Inc
2121 Rebild Drive
Solvang, CA 93463

Algebra

| 12 | Identifies complex symbols: parentheses, brackets, intersections, etc. |

Have You Got a Match?
Grade Level: Middle

MATERIALS: Story-problem matching page (Appendix G), scissors

ORGANIZATION: Cooperative groups of two, three, or four

PROCEDURE: This activity will give students practice in reading two- and three-step story problems and identifying the equation which might be used to solve each problem.

Give each group a copy of the story-problem matching page and have them cut out the equation cards. Working together in groups, have them read each problem and find the matching equation. Upon finding the matches, they must also explain why parentheses and brackets were used in each problem.

Once they have successfully completed this portion of the lesson, have each group write three, word problems and matching equations and pass them from group to group for others to solve.

1. Juan, Sharon, and Andy were playing a game of darts. Juan made a score of 3 points, Sharon scored two times as many points as Juan. Andy scored one more than Sharon. How may points did Andy score?

$(2 \times 3) + 1 = a$

2. The Eighth grade students were helping the kindergarten students in the computer lab. There were five eighth grade students helping. There were four times as many kindergarteners as eighth graders. How many students were in the computer lab?

$(4 \times 5) + 5 = s$

3. Tony agreed to bake 125 cookies for the class party. He baked 42 cookies on Wednesday and 36 cookies on Thursday. How many more cookies were left to bake?

$(125 - 42) - 36 = c$

Hands On, Inc
2121 Rebild Drive
Solvang, CA 93463

Algebra

| 13 | Identifies and uses the commutative property |

It's All Bean Said Before
Grade Level: Middle/Upper

MATERIALS: Red beans, blue beans (or pinto and kidney beans), unlined paper, pencils

ORGANIZATION: Teams of two students

PROCEDURE: The commutative property of addition states: The order of TWO addends does not affect the sum. This activity is designed to have students prove and explain this property.

Give each student four red beans and two blue beans. Explain that the group of red beans will be called "a" and the group of blue beans will be called "b." On the chalkboard write a + b. Have the students place four red beans on a piece of paper and write in an addition sign and then add the two blue beans.

Next, write the second part of the property on the board so it reads a + b = b + a. Inquire how many blue beans are on the left side of the equal sign? How many red beans? How many will you need on the right side? How many blue beans will you need on the right side if they are equal? Have students set up the beans following the a + b = b + a pattern from the board on their paper.

Once students are able to express that the position has changed, introduce the word "commute" which means to change position. Tell students that addition has a rule called the "commutative property" which means that you can change the order in which you add numbers without changing the outcome.

Have students experiment with other combinations and have them verbalize the commutative property to fellow group members.

Hands On, Inc
2121 Rebild Drive
Solvang, CA 9346:

Algebra

| 13 | Identifies and uses the commutative property |

Element..ary My Dear Bean
Grade Level: Middle/Upper

MATERIALS: An assortment of beans

ORGANIZATION: Individually or in pairs

PROCEDURE: The commutative property of addition says that the order in which two numbers are added does not affect the sum of those two numbers. In this activity students will experiment with chemistry (using beans instead of atoms) to reach an understanding of this property.

Tell your class that you are going to pretend that you are a scientist who has discovered a new compound which you are going to call "beanoleum (B)." Hand out a pile of beans with a good mixture of the different varieties. Tell them what they have is a lump of raw beanoleum ore.

Select two types of beans to represent the elements in beanoleum. We'll label these elements S and C (S will be kidney beans, C will be split peas). Tell your students that you have yet to find out what the chemical spelling of beanoleum is, but you know that in three molecules there are 15 atoms of C. Have your students separate fifteen split peas from their "lump of ore."

Next, say that in four molecules of B, there are eight atoms of S. Have them separate eight kidney beans from the pile and have teams create a chart to represent the information they know (see sample).

From this information, they should complete the chart and write equations. Emphasize that the order in which the atoms are mixed is inconsequential. The end result will always be Beanoleum. Have students verbalize or write paragraphs explaining how the creation of Beanoleum proves that addition is commutative. Have them repeat this process with new compounds they create from their bean pile.

Number of Molecules	S	C
1		
2		
3		15
4	8	

Hands On, Inc
2121 Rebild Drive
Solvang, CA 93463

Algebra

| 13 | Identifies and uses the commutative property |

Hip, Hip Array
Grade Level: Middle/Upper

MATERIALS: Paper clips and grid paper

ORGANIZATION: Whole class activity eventually dividing into groups of two

PROCEDURE: This activity will provide a manipulative process for understanding the COMMUTATIVE PROPERTY OF MULTIPLICATION.

First clarify the terms "row" and "column" with the students so they understand the difference. Then explain that today they will call the rows "a" and the columns "b". On the chalkboard write the first part of the commutative property, ab. Reinforce the concept that two variables placed side by side mean multiplication.

Give each student twelve paper clips. Then write a = 4 and b = 3. Have the students place their paper clips in the appropriate pattern on their desks. Monitor for correct pattern. Next, write the second part of the commutative property ab = ba. Have the students rearrange the paper clips to fit this part of the pattern.

Ask if the two patterns are the same. Have the students verbalize their likenesses and differences. Explain that the commutative property says that the order in which you multiply does not affect the total amount.

Have the students get in their groups. Have the first student use paper clips to demonstrate the first part of the property and the second student demonstrate the second part of the property. Repeat the procedure with varying amounts of paper clips. The first student should verbalize the amount of "a" and "b", as should the second student. The roles should be switched so each student has practice in each role.

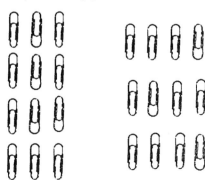

Hands On, Inc
2121 Rebild Drive
Solvang, CA 9346:

Algebra

| 13 | Identifies and uses the commutative property |

Staring at Stairs
Grade Level: Middle

MATERIALS: Paper, pencils and a marker or chip of some type

ORGANIZATION: Whole class activity then teams of two

PROCEDURE: We tell students that addition and multiplication are commutative and show them why. We need to also show them why subtraction and division are NOT commutative. This is the purpose of this lesson.

Begin with a whole class explanation of the commutative property and have each student draw a stairway as shown below. On the overhead, write "3 + 1" and have the students move their markers up three steps and then up one step. Ask what step they are on. Repeat the procedure with "1 + 3." Elicit from students that both combinations end up on the fourth step.

Next, write "2 x 3" and explain that this means 2 sets of 3 steps. Have them move their markers in this manner. Do "3 x 2" and point out the these two combinations also end up on the same step. At this point you should reintroduce the concept of the commutative property.

At this point, divide the class into groups of two and pose the questions: "Is subtraction commutative?" and "Is division commutative?" Tell students that they must work together to answer these questions and they must prove their answers by using the stairs and markers.

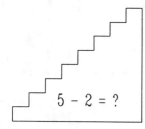 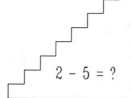 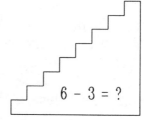 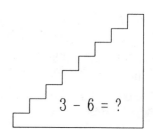

Algebra

| 14 | Identifies and uses the associative property |

Round and Round You Go
Grade Level: Middle/Upper

MATERIALS: Duplicated pages of road maps, paper and pencils

ORGANIZATION: Individually or in groups of two

PROCEDURE: The associative property of addition says that when three or more numbers are added, the order in which the numbers are added does not affect the sum. Students need concrete experiences to help them understand and remember the associative property. This activity will provide such an experience.

Give students copies of a road map and discuss the various features, such as how to read mileage, how to identify highway numbers, how to select various routes to arrive at the same destination, etc.

Tell students that they are going to plan a short outing in which they will be driving in a circle (loop). Have them search on their maps to find a loop that they would like to drive. Their task will be to figure the number of miles they will drive to complete the loop. They will then write a number sentence (equation) to describe this drive. The easiest way to do this is to assign a letter symbol to the distance between each town.

The next step is to have students imagine that they are now driving the loop in the opposite direction. They should once again figure the mileage and write an equation for this direction. They will argue that it is silly to do this because the mileage will be the same, which is exactly the point of the lesson (and the associative property). It doesn't matter what order the numbers are in when we add, the answer will always be the same. Going through this process will provide a mnemonic device (of sorts) which students will remember. The next time you begin discussing the associative property by saying, "Remember the map?"

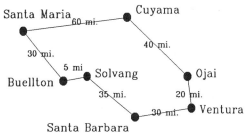

$35 + 30 + 20 + 40 + 60 + 30 + 5 = 220$

$5 + 30 + 60 + 40 + 20 + 30 + 35 = 220$

Hands On, Inc
2121 Rebild Drive
Solvang, CA 9346?

Algebra

| 14 | Identifies and uses the associative property |

What Goes Where, When?
Grade Level: Middle

MATERIALS: Various types of fruits to make a fruit salad, bowls, utensils for cutting and eating

ORGANIZATION: Whole class or cooperative groups of four

PROCEDURE: This lesson uses the students' inherent interest in cooking and eating to teach the associative property of addition.

On the day before the lesson, talk with the class about the ingredients of a good fruit salad. Through this discussion, have the students narrow their choices to three or four ingredients and have them decide upon the number of pieces (four watermelon cubes, two peach slices, etc.) they wish to include for their personal salad.

On lesson day, have various groups be responsible for slicing/chopping/cutting the fruit into appropriately sized pieces. All fruit types should be kept in separate bowls after cutting. Bring the entire class back together to do the mixing of ingredients. Write the formula for the ingredients on the board: 3a(pple)+4b(anana)+5w(atermelon) = s(alad) and mix these ingredients together in one bowl.

Now, tell students that you are going to make a different kind of salad and write this formula on the board: 4b+5w+3a=s. Someone will undoubtedly point out that this is the same salad, which is your cue to write the following equation on the board: 3a+4b+5w=4b+5w+3a and ask students if this equation is true. Tell them that this is an example of the associative property of addition. Discuss other possible combinations.

Let students finish the mixing of fruits for their salads and then enjoy the "fruits" of your teaching.

Algebra

| 14 | Identifies and uses the associative property |

Enjoying the Easy Times
Grade Level: Middle/Upper

MATERIALS: Small boxes, slips of paper, and pencils

ORGANIZATION: Groups of two

PROCEDURE: The associative property of multiplication says that when three or more numbers are multiplied, the order in which they are multiplied does not affect the product. This activity will provide a manipulative approach to understanding this property.

The teacher should begin by having two students write the numbers 2,3,4,5,6,7,8,9 on separate small pieces of paper and place them in a small box.

Divide the class into teams of two and have one team member draw one number out of the box and record it on paper as "a = ." The same person should draw another number and record it as "b = ," Draw a third number and record as "c = ."

During this time, the teacher should write the associative pattern on the board as "a(bc) = b(ca) = c(ab). Have each team record their combinations on paper, following the pattern of the associative property. For example: 5 (2 X 6) = 2 (6 X 5) = 6 (2 x 5).

Ask students which combination would be the easiest to do without using paper and pencil. Which would be the hardest? Emphasize the fact that the order in which the numbers are multiplied does not change the total amount.

Next, have the other member of the pair draw the numbers from the box while his partner writes the possible combinations of the associative property. Repeat this procedure several times while emphasizing the usefulness of choosing particular numbers to multiply first.

Hands On, Inc
2121 Rebild Drive
Solvang, CA 9346

Algebra

| 14 | Identifies and uses the associative property |

Multiplying Darts
Grade Level: Middle/Upper

MATERIALS: A dartboard for suction cup darts or bean bag toss game; a bullseye might even be drawn on the board with students throwing a small damp sponge

ORGANIZATION: A whole class activity or small groups of four to six students

PROCEDURE: Most students have played bean bag toss games or darts in which the score is figured by totalling the amount of three "tosses." In this game, students will figure their total by multiplying the amount of each of the three tosses. The person, or team which does the multiplication most quickly is the winner of those points.

Begin by throwing three darts at the board. For example, imagine that the three numbers are 5, 7, and 12. Ask students to multiply these three numbers together. Many students will immediately respond, "5 x 7 is 35" but have then placed themselves in a position of having to multiply two 2 digit numbers to find the final answer. Point out that by first multiplying 7 x 12 = 84 and then multiplying 84 x 5, it would enable them to do the second computation much more quickly.

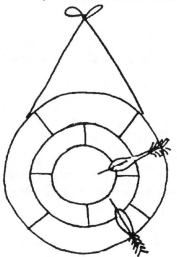

This technique of selecting two of three given numbers to multiply first and then multiplying the product by the third number is making use of the associative property of multiplication.

Do one or two more sample problems and then divide the class into teams of students, or select three students to throw darts at the same time. Our experience with this lesson shows that with three students throwing darts at the same time, students are forced to choose two numbers to multiply. If darts are thrown one by one, students will multiply each number as it is thrown.

Hands On, Inc
2121 Rebild Drive
Solvang, CA 93463

Algebra

| 15 | Identifies and uses the distributive property |

What's Cookin?
Grade Level: Middle/Upper

MATERIALS: A copy of a recipe from home for each student, beans or blocks representing each item in the recipe

ORGANIZATION: Individually

PROCEDURE: The distributive property says that the product of two numbers can be expressed as the sum of two products. In other words, a whole can be broken down into parts, multiplied, and then regrouped. In this activity, students will distribute the ingredients of a recipe to make the recipe larger or smaller.

Have students bring a recipe which contains various measured ingredients. Explain that in each recipe, the amount of each ingredient must be exactly right if the dish is to taste right. You can use beans or blocks to represent the different recipe items and have students make a mock-up of their dish.

Have students rewrite their recipes using letters to represent the ingredients. They now have an equation using variables. An example might be:

6 cup of flour	6f
2 1/4 cups of milk,	2 1/4m
1 packet of yeast	y
1 Tbls butter	b
2 tsp salt	2s
2 Tbls sugar	2g

$$6f + 2\ 1/4m + y + b + 2s + 2g = b(bread)$$

In most recipes, there are two or more steps. It is rare when all ingredients are lumped together before any mixing takes place. When making bread, the yeast and flour are mixed first (multiplied). In a separate pan, the milk, sugar, butter and salt are combined (multiplied). Then, the two sets of ingredients are combined in a mixing bowl.

After students have simulated this process, using beans as their ingredients, have them transfer their knowledge to numbers. Give them any three digit number and have them break it down into three or four separate, smaller numbers (like the different bowls of ingredients). The numbers when combined will then add up to the original total, regardless of the way in which they were divided into groups.

Hands On, Inc
2121 Rebild Drive
Solvang, CA 9346:

Algebra

| 15 | Identifies and uses the distributive property |

Here's a Tip For You!
Grade Level: Upper

MATERIALS: Menus from a restaurant, sets of coins and bills (Appendix E)

ORGANIZATION: Groups of two

PROCEDURE: This activity will show how to use the distributive property in everyday life when pen and paper are not available.

Begin by giving a copy of the menu to each student. Give them some time to write down their order for lunch, dinner, or whatever. Have them figure their total bill.

Next, ask each student to figure a 15% tip for the waiter. At this point, you will find that many students have difficulty and will want to figure this by writing it down. Explain that it is rare for them to have paper and pencil in a restaurant and they will need to figure this "in their heads."

Explain that in order to figure the proper tip, students can distribute the 15% into 10% + 5% and eliminate the need for paper and pencil. Have them figure 10% of the total bill and add another half of that amount (5%) giving the proper tip. You should also remind them that rounding the original price makes this process easier.

Give students play money and let them order meals and then practice leaving the proper tip.

Example:
15% of $5.95 = TIP
(10% x $5.95) + (5% x $5.95) = TIP
$.60 + $.30 = $.90

Hands On, Inc
2121 Rebild Drive
Solvang, CA 93463

Algebra

| 15 | Identifies and uses the distributive property |

Distributing Discounts
Grade Level: Upper

MATERIALS: Newspapers with sale advertisements, glue, construction paper

ORGANIZATION: Whole class activity or teams of two

PROCEDURE: In this activity, students will use the distributive property of multiplication to figure discounts on various sale items. The emphasis will be on doing computations "in their heads" rather than with paper and pencil.

Hand out a newspaper to each student and have them go through and cut out advertisements with reference to percentage discounts. Have them glue each advertisement on a separate piece of paper, and on each piece of paper, have them list a particular item which might be sold at that store and its list price.

At this point, coach the students in the process of rounding off the original price, figuring 10% of the total, and then doubling, tripling, halving, etc. this amount to arrive upon the proper discount.

For example if a camera's list price were $219.00 and a 25% discount were offered, round the $219.00 to $220.00, figure 10% ($22.00) and then add $22.00 + $22.00 + $11.00 (the remaining 5%). 220 − 55 = $165.00

Student papers should be circulated around the classroom giving each student numerous opportunities to estimate the sale cost of these items.

Algebra

| 15 | Identifies and uses the distributive property |

Mental Math
Grade Level: Middle/Upper

MATERIALS: Scissors, graph paper (1 cm), chart paper

ORGANIZATION: Whole class activity

PROCEDURE: The purpose of this lesson is to help students learn to use the distributive property to find a solution without using pencil and paper.

Distribute graph paper to each student and explain that they will cut the graph paper into units of tens and ones. They will then use these units to help solve multiplication problems. Begin with the following example: 3 x 17 =

> 17 = 1 unit of 10, AND 7 units of 1
> therefore,
> 3 x 17 = 3 units of 10, AND 21 units of 1
> which can easily be added without pencil and paper

As you explain this procedure, have the students manipulate their graph paper pieces to "picture" the process of distributing numbers.

Provide several sample problems and then let students pair up to explain the process to one another.

10   7

Algebra

| 16 | Identifies extraneous or missing information in word problems |

If I had
2 parakeets
and bought
3 owls

Word Wary
Grade Level: Middle/Upper

How many nocturnal animals would I have?

MATERIALS: Markers, sentence strip cards

ORGANIZATION: Whole class activity

PROCEDURE: In this teacher directed lesson the students willl be acting as if they are part of a word problem. They will hold up signs such as, "What is the difference between?" "How much bigger?" or "How many?" While other students will have signs such as, "three rabbits," "four parakeets," "two birds," etc.

Begin by giving each student a sentence strip card, and direct the students to write a certain number and then a noun which that number might describe. An example might be, "two parakeets." You may want to set the number range so students are working with the types of numbers that they can handle.

Select two students to go up to the front of the room, carrying their cards with them. In our example we'll use THREE OWLS and TWO PARAKEETS. Ask the students if they can create a word problem involving owls and parakeets. As they make up a problem, point out the types of words or phrases that are used between the numbers in a word problem, words such as "how many?" "how much more?" "is equal to," "in all," etc. Have each student write one such phrase on the back of the sentence strip.

Select students with apropos statements to come to the front of the room with their phrases, standing with the two original "number" students, to display a human word problem.

Continue by having other students come forward with their phrases. Show students how different phrases change the operation and order of the problem. You'll find students will begin to arrive upon some humorous additions including problems which contain extraneous or missing information.

Hands On, Inc
2121 Rebild Drive
Solvang, CA 93463

107

| 16 | Identifies extraneous or missing information in word problems |

Finding the Rhyme and Reason
Grade Level: Middle/Upper

MATERIALS: An assortment of Mother Goose nursery rhymes, such as "Jack-Be-Nimble," "The Old Lady Who Lived in a Shoe," and "Wee Willy Winkie"

ORGANIZATION: Individually or in teams of two.

PROCEDURE: This is a fun activity in which students will be creating word problems based upon nursery rhymes. In our example, we will use Hey Diddle Diddle.

Hey Diddle Diddle, might be rephrased in a word problem to say, "Once there was a cat who played the fiddle to accompany his business partner, a moon-jumping cow. The cat also had a friend who was a dog and happened to have a wonderful sense of humor. This dog also had an enormous appetite and was in the constant company of a bowl and spoon. How many objects, in all, would be included in this story? How many living objects are there in this story?

This will obviously take some creativity on your students' behalf and therefore, you may choose to do this in teams of two. Show students that in the problems that they create, they should incorporate information that is missing, or information that is extraneous.

An example involving missing information might be: if Humpty Dumpty sat on a wall and fell. If all the king's horses and all the king's men couldn't put him together again, how many horses and men were there in all?

Let your students share their creations with the entire class.

Algebra

16	Identifies extraneous or missing information in word problems

Checking the Stats
Grade Level: Middle/Upper

MATERIALS: Sports sections from several days papers

ORGANIZATION: Cooperative groups of four students

PROCEDURE: In this lesson, the students will use statistics from the sports section to create word problems with either missing or extraneous information.

The teacher should begin the lesson by displaying a set of won-loss records from one of the major leagues and create some sample problems such as: If Buffalo has won nine games and lost two, and if Green Bay has won four games, how many games has Green Bay lost if both teams have played the same number of games? Or, If Washington has lost three games and has played two more games than Buffalo, how many games has Washington won?

Give each cooperative group a set of sports statistics and have them write a set of word problems which can be exchanged with other groups. Once students have mastered this portion of the lesson, have them write a separate set of problems, but this time they should add extraneous or unnecessary information. Rotate the sets of problems around the room for each group to solve.

A sample problem with extraneous information might be: Denver has scored 67 points in five games, Kansas City has scored 43 points in five games, and if Seattle has scored 6 touchdowns in their five games, how many more points per game has the Colorado team scored than Los Angeles?

AFC WEST STANDINGS				
	WON	LOST	POINTS FOR	POINTS AGAINST
DENVER	5	0	67	34
SEATTLE	4	1	54	42
LOS ANGELES	3	2	54	33
KANSAS CITY	3	2	35	41
SAN DIEGO	1	4	23	61

Hands On, Inc
2121 Rebild Drive
Solvang, CA 93463

Algebra

| 16 | Identifies extraneous or missing information in word problems |

Extra Parts
Grade Level: Middle/Upper

MATERIALS: Magazines and newpapers, scissors, and paper

ORGANIZATION: Individually or pairs

PROCEDURE: Explain to your students that in order to figure out the solution to a problem, it is often just as important to identify which information is extraneous or missing as well as that which is necessary.

In this activity, students should cut out five miscellaneous pictures from magazines. Next they should cut those pictures into component parts (a photo of a car gets separated into the tires, doors, etc.; a photo of a person gets separated into arms, legs, etc.). This should result in each student having a pile of various parts to various objects. They should then draw two circles on their paper and label them EXTRANEOUS INFORMATION, and NECESSARY INFORMATION.

At this point, they should think of a word which might be represented by a conglomeration of pictures (see below). They (player 1) should display all cutouts in random order and ask a partner (player 2) to discover the secret word. To do this, player 2 should point to a cut-out. Player 1 then places this selected picture into the appropriate circle -- Extraneous or Necessary. Once enough pictures have been separated, the partner should make an educated guess as to what the mystery object is.

It should be pointed out to your students that the purpose of this assignment is not necessarily to arrive at the "right" answer, but to come up with a reasonable guess based on the information obtained from separating the "extraneous" information from the "necessary."

A sample display might include

a bicycle (wheels, frame, handlebars),
a kitchen sink (drains, faucet, basin),
a tree (trunk, leaves, branches),
a table (legs, table top),
a clock (hands, numbers, face)

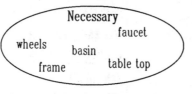
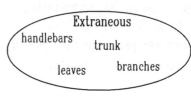

The mystery item is a motor home

Hands On, Inc
2121 Rebild Drive
Solvang, CA 93463

Algebra

| 17 | Identifies open and closed number sentences |

To Loop or Not to Loop
Grade Level: Middle/Upper

MATERIALS: Highway road maps for each team

ORGANIZATION: Groups of two students

PROCEDURE: In this activity we will use a "loose" translation of the concept of open and closed number sentences. In this case, if there is only one way of moving from point A to point B, it will be considered a closed statement. If there are a variety of ways to get from one point to the next, it is an open statement.

Make an overhead projection of one section of one of the highway maps or post a section of the map which is big enough for all students to see. Find two towns which lie on one highway route and ask students if there is any other logical way of traveling the given road to get from town 1 to town 2. You may find that there are small roads branching off that wind around and connect the two towns and in this case, there would be alternative routes. However, you may find that there are no substitute roads that leave the main highway.

If there is only one connection between 1 and 2, then the route will be called a <u>closed</u> route. If there is more than one way of travelling between towns, then it is an <u>open</u> route.

Once students understand the definitions, have them go through their maps and find examples of open and closed routes. Have them find cases in which there are two, three or four routes.

As an extension, you may wish to have students write equations for the mileage between the towns. You may find that some students get carried away with huge, circuitous routes.

The underlying purpose of this lesson is to have students identify that a closed sentence is "cut and dried." There are no variables and no alternate solutions.

Hands On, Inc
2121 Rebild Drive
Solvang, CA 93463

Algebra

17 | Identifies open and closed number sentences

An Indefinite Answer
Grade Level: Middle/Upper

MATERIALS: No special materials, but students must have a knowledge of the parts of speech

ORGANIZATION: A whole class activity

PROCEDURE: This is a math lesson which uses English grammar as an approach to teaching the concept of open and closed sentences.

Write the following sentence on the board to begin the lesson.

--- boys and --- girls filled --- spaces at the table

Ask a student to come forward and write a word or number in the three blank spaces. Some sample responses might be:

1. Three boys and four girls filled seven spaces at the table.
2. Ten boys and five girls filled eight spaces at the table.
3. Some boys and a few girls filled many spaces at the table.
4. Tall boys and short girls filled six places at the table.

Have several students come forward to add their words until one of them puts numbers in the blanks. At this point, ask the class if any of the sample sentences are true or false. The point to identify is that the only sentences that can be judged true or false are those sentences that have a numerical or "fixed value."

In the examples above, sentence 1 is closed because it can be proven; sentence 2 is closed even though it is false because it too can be proven. Sentences 3 and 4 MAY be true but cannot be proven, therefore they are open number sentences. The defining feature is, "Can the sentence be proven to be true or false given the information provided?"

Once students understand the concept, have them write several open and closed sentences. You can also demonstrate to students the reasoning behind the term "indefinite pronouns." This term is supported by the fact that the information they give is non-specific and hence, indefinite.

Hands On, Inc
2121 Rebild Drive
Solvang, CA 93463

Algebra

| 17 | Identifies open and closed number sentences |

Missing Jigsaw
Grade Level: Middle/Upper

MATERIALS: Magazines and newpapers, glue, scissors, poster paper (medium size)

ORGANIZATION: Individually, and then in pairs

PROCEDURE: In this activity students will be developing their ability to identify missing information as it applies to open and closed sentences.

Start by having students cut out various pictures from periodicals. Each students should select three pictures to work with. After selecting a picture, they should glue it to a piece of posterboard and then cut the picture into eight to fifteen jigsaw-type pieces.

Divide the class into teams of two and have one student in each group randomly display all of the jigsaw pieces for one of the pictures. The student can secretly choose to display all puzzle parts or can hide one or two pieces from the other player. The second student should carefully look at the pieces, without touching them, to see if all of the pieces are there. If student 2 feels that all puzzle parts are present, he says, "CLOSED," meaning that all information is present. If student 2 feels that a piece is missing, he says "OPEN," to indicate that there is not enough "information" (jigsaw pieces) to complete the puzzle.

If the response is correct, player 2 gets 5 points. If student 2 decides that by looking at the original set of puzzle parts, he cannot decide whether it is open or closed, he may move pieces around, BUT loses one of his five possible points for each puzzle piece that he touches. If his guess is incorrect, player 1 picks up the puzzle parts and the game begins over with player 2 displaying his jigsaw pieces.

The game goes back and forth until each player has displayed his/her three puzzles. The student with the most points wins. Players can then switch teams and repeat the process.

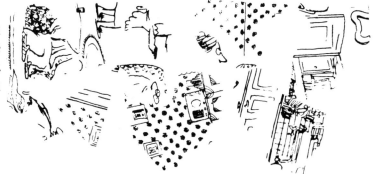

Hands On, Inc
2121 Rebild Drive
Solvang, CA 93463

Algebra

17	Identifies open and closed number sentences

Closing Out the Game
Grade Level: Middle/Upper

MATERIALS: Hando boards (Appendix A), cards with numbers and letters

ORGANIZATION: Whole class

PROCEDURE: The object of this game is to identify as many closed number sentences as possible. The teacher should begin by reviewing closed and open number sentences.

Begin the game by giving each student a Hando board and having them make a set of letter cards -- one card for each number, 1 - 15. Have them write an operation (+ or -) sign in each space under the A on their HANDO boards and and equal, greater than, and less than sign in each space under the D. Each student and the caller should have a set of number cards.

As the caller draws a card and calls a number, the students place the number in any space they wish. As the game progresses, the students may move the numbers to any other space they want. The idea is to rearrange them into closed number sentences so as to identify as many closed number sentences as possible.

The winner is the student who has created the most closed number sentences on the board. The winner may cover the whole board or only those spaces which make closed sentences.

The "free space" should be covered with a number card

H	A	N	D	O
	+		=	
	−		>	
	+		=	
	−		<	
	+		=	

Hands On, Inc
2121 Rebild Drive
Solvang, CA 93463

Algebra

18	Applies order to operations: multiplication, division, addition, subtraction

Placing an Order
Grade Level: Middle/Upper

MATERIALS: Overhead projector or chalkboard

ORGANIZATION: Cooperative groups of two to four students

PROCEDURE: This is an early exercise in practicing the order of operations. Begin by explaining to students that operations must be done in a specific order in algebra problems. The order is multiplication, division, addition, and subtraction. A mnemonic device for memorizing this is: My Dear Aunt Sally.

In this activity, a group of numbers will be written across the screen or chalkboard. The object of the game is to strategically use the types of operations given by the teacher in an order that will create the highest possible total once the equation is solved.

An example might be:
    Teacher writes   4   6   8   9   3   5   and the symbols +, +, −, x, x

    Given this, student A might write 4 x 8 + 9 − 3 + 5 x 6 =
        Which simplifies to: 32 + 9 − 3 + 30 = (multiplication first)
        Which simplifies to: 41 − 33 = (no division, so addition next)
        Which gives a final total of 8 (subtraction last)

    Student B might write 8 x 9 x 6 +5 + 4 − 3 =
        Which simplifies to: 432 + 5 + 4 − 3 =
        Which simplifies to: 441 − 3
        Which gives a final total of: 438

Write several examples on the board and divide the class into groups of two or four to complete this activity.

Hands On, Inc
2121 Rebild Drive
Solvang, CA 93463

Algebra

18 | Applies order to operations: multiplication, division, addition, subtraction

I'll Even Give You the Answer...
Grade Level: Middle/Upper

MATERIALS: Calculator, chart with the order of operations for equations, stopwatches or second hands on student watches

ORGANIZATION: Teams of four to six students

PROCEDURE: This is an early lesson in practicing the order of operations for equations: Multiply, Divide, Add, and Subtract or (M.D.A.S. My Dear Aunt Sally).

The teacher should begin by writing a number on the board (we'll use 52 as an example). The students must then select numbers to first multiply, then divide, then add and subtract to equal the given number. For example: given 52 as the desired total, the students begin with 100 and then...

Multiply	100 x 2
Divide	200 − 4
Addition	50 + 6
Subtraction	56 − 4
Equals	= 52

Next, the teacher should explain to the students that they will be working with their team member(s) to solve a problem using the four steps in the order of operations. This work must be done on paper and saved to show to the opposing team.

Opposing team members will time the first team and then check for the four steps and correct order of those steps. The winning team will have taken the shortest amount of time to solve the problem. The teacher should supply the number for each set of operations.

As an extension, this lesson can become very complex, including such extras as parentheses and brackets or negative and possitive numbers. Also, student teams could create the solution numbers.

Hands On, Inc
2121 Rebild Drive
Solvang, CA 93463

Algebra

| 18 | Applies order to operations: multiplication, division, addition, subtraction |

Around and Around
Grade Level: Middle/Upper

MATERIALS: Small cards with numbers written on them (two cards for each number 1 through 12), box, score sheets, calculators if desired

ORGANIZATION: Cooperative groups of three

PROCEDURE: The teacher should explain MDAS see provious lesson) to the class and say that they are going to play a game to practice this.

Place the cards in a box and give a box to each group of three along with a score sheet for each child. Group members should draw a number to see who goes first. High number starts the game by drawing a slip and laying it down, face up. The second player draws a number and <u>multiplies</u> it by the first number that is on the table. The answer must be written on that player's score sheet.

The third player draws a card and must <u>divide</u> the answer on player two's score sheet by this number. If he is unable to do the division, he may draw another card. If he is still unable to divide by this number, he loses his turn. If he can do the division, he must write the answer on his score sheet.

The first player draws a slip and <u>adds</u> the number to the previous answer on player two or three's score (the one with the last correct answer). Player one should write the answer on his score sheet.

Player two then draws a card and <u>subtracts</u> the number from player one's answer and writes it on his score sheet. The play keeps going from one player to the next, each using the function in its turn (MDAS) until all the slips have been drawn. Each player then adds the scores that he has computed and high score wins.

Player 1	Player 2	Player 3	Player 1	Player 2
Draws a 3	Draws a 6 and multiplies	Draws a 4 Unable to divide Draws a 3 and divides 18 by 3	Draws a 10 adds to 6	Draws a 1 and subtracts from. 16
	18	6	16	18+15
	writes 18 on score sheet	writes 6 on score sheet	writes 16 on score sheet	writes 15 on score sheet

Hands On, Inc
2121 Rebild Drive
Solvang, CA 93463

Algebra

| 18 | Applies order to operations: multiplication, division, addition, subtraction |

Valuable Letters
Grade Level: Upper

MATERIALS: None, although students might cut out letters and operations signs

ORGANIZATION: Cooperative groups of four

PROCEDURE: In a complex math sentence or equation the four basic operations are performed in the following order: multiplication, division, addition, subtraction. However, students must also be aware that parentheses and brackets also indicate special treatment. This lesson combines these concepts.

Assign each consonant a value and each vowel an operation, since there are five vowels and four operations, one operation must be represented by more than one vowel. Students are to calculate the "value" of their names. They will need to use parentheses to separate the different operations.

For example, if each consonant has a sequential value starting at one, the letters in HANDS ON would be represented by 6 = H, 11 = N, 3 = D, 15 = S, 11 = N; if the vowels a, e, i, o, u are multiplication, division, addition, subtraction, and multiplication then the equation for Hands On would be (6 x 11) 3(15 − 11) = ? Notice that there is no operation between the 3 and the two groups of numbers in parentheses. You should let students wrestle mentally with what to do with this situation.

One solution is to point out that in math, especially in algebra, when an operation is not indicated it is assumed that the numbers should be multiplied. Students will also be able to <u>simplify</u> their name equations. The equation for HANDS ON could be simplified as follows: (6 X 11) 3 (15 − 11) = 3(6 x 11)(15 − 11). Point out that regardless of MDAS, the operations within parentheses must be done first.

An extension could be to see what is the "most expensive" food, vehicle, or which class has the highest scoring collection of names.

Hands On, Inc
2121 Rebild Drive
Solvang, CA 93463

Algebra

| 19 | Uses number lines to solve equations with positive and negative numbers |

AD/BC
Grade Level: Middle/Upper

MATERIALS: String to suspend across the room, measuring tape, masking tape

ORGANIZATION: Whole class or small group activity

PROCEDURE: The labels A.D. (anno domini) and B.C. (before Christ) are familiar to most middle grade children. Working with and understanding these abbreviations is another matter. Students have difficulty understanding that B.C. starts at 1 and moves backwards into time as the numbers get bigger. In a way, it is comparable to the relationships of positive and negative numbers.

Begin the lesson by suspending a string as a number line, across the classroom and measuring equal segments (perhaps every foot or 30 cm). Mark each segment with tape as follows: select a central point and label it 0. Moving out from this point in each direction, label segments in increments 100, 200, 300, etc.

Discuss with your class the concept of A.D. and B.C. and show them how this is similar to positive vs. negative numbers. Tell them that their project for the day will be to find "midpoints" between two dates. Have one student stand at 150 A.D. and another stand at 300 B.C. Ask the class to estimate the location and the exact date of the midpoint of these two dates and explain that when they add negative and positve numbers, this is essentially what they are doing -- finding the average between the given numbers. This concept is difficult and will need a lot of discussion and many samples.

The point of the lesson is not necessarily to have students learn to add negative and positive integers; more than that, this will give students a way to relate to WHY negative and positive numbers are treated in special ways.

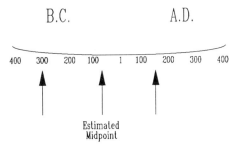

Continue the discussion by eliciting statements about years and integers.

Algebra

| 19 | Uses number lines to solve equations with positive and negative numbers |

Adding Nothing?
Grade Level: Middle/Upper

MATERIALS: Cups, beans, number lines

ORGANIZATION: Individually or in teams of two

PROCEDURE: One concept which students have difficulty grasping is that numbers can be added and the result is zero or a quantity less than originally added. This situation takes place when adding positive and negative numbers. This activity demonstrates to students why this occurs.

Give each students two cups and twenty to thirty beans. Label one cup + and one cup −. Next, have them draw or label a number line as shown. The number line must have sufficient space between numbers so the cup base can fit.

Have the students place the − cup on −5 and place five beans in the cup, have them place the + cup on +5 and put five beans in. Write the following equation on the board: −5 + 5 = ? and ask for possible solutions. They will range from 0 to −10 to +10. Tell them that the correct answer is 0 and then see if they can explain why this is true.

The explanation should involve removing one positive and one negative bean at a time -- in other words, "cancelling one another out" since they have opposite value on the number line.

To give students more practice, write some other equations on the board such as: +5 + ⁻3, or +3 + ⁻8.

Physically removing beans and seeing the relationship on the number line will help students to understand this concept.

Algebra

| 19 | Uses number lines to solve equations with positive and negative numbers |

Double Negatives
Grade Level: Middle/Upper

MATERIALS: No special materials are necessary

ORGANIZATION: Whole class activity

PROCEDURE: When working with equations, students will often have to deal with negative numbers which must be changed to positive. By using double negatives in sentences, students will see how two negatives make a positive.

Begin by generating a list of negative words with the class which should include, never, none, no one, no, not, and nothing. Ask students to explain why these words are considered to be negative, eliciting that they make a positve statement into the opposite.

Write, "I am going to the store," on the chalkboard and ask what makes this sentence positive -- the word "am." See if anyone can think of another word to add which makes this sentence even more positive. Some examples might be, "definitely," "always," or "certainly." Show students that by using these words, the sentence remains positive -- in other words, two positives added together make a positive.

Now add a negative to the sentence, "I am never going to the store." Point out that the positive word "am" is still there but the one negative, "never," makes the sentence negative. In other words a positive and a negative combined make a negative. Even if the positive word "certainly" is added, "I am certainly never going to the store," it remains negative.

Now add a second negative, "I am not never going to the store," and show students that if someone is "not" "never" going to the store, then the sentence has once again become positive -- they are going to the store. At this point explain that two negatives in a row "cancel one another out" and become positive. Have students experiment with writing sentences demonstrating this concepts.

Hands On, Inc
2121 Rebild Drive
Solvang, CA 93463

Algebra

| 19 | Uses number lines to solve equations with positive and negative numbers |

Submarine Fuelishness
Grade Level: Middle/Upper

MATERIALS: Graph paper

ORGANIZATION: Individually

PROCEDURE: In this exercise students will practice working with positive and negative numbers by steering a mini-sub through an underwater cave city, and then computing the amount of fuel used in the voyage.

They will be using compass headings to represent the directions: n(north) is up, e(east) is right, s(south) is down, and w(west) is left. No other movements are possible. Since the sub will float to the surface without the use of power, each square moved N will = 0. All other directions will be considered negatives: e = −1, w = −2, and s = −3.

Have each student construct an underwater cave in the middle of the graph paper. The cave should be drawn using right angles only but include at least one turn in each direction.

Have students exchange the cave maps and have them record the sub's movements on a separate piece of paper in the form of an equation. Upon escaping the maze, have them compute the amount of fuel consumed figuring that each single square on the graph paper represents one gallon.

A sample might be: 4n + 3e + 6s + 2e + 1s + 4e
(4 x 0) + (3 x −1) + (6 x −3) + (2 x −1) + (1 x −3) + (4 x −1) =

The answer to the equation will be the number of gallons used shown as a negative integer. As an extension, you can change the value of each direction to a combination of positive and negative numbers or construct more complicated grids and have students attempt to escape using the least amount of fuel.

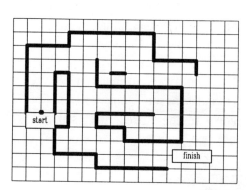

Hands On, Inc
2121 Rebild Drive
Solvang, CA 93463

Algebra

| 20 | Uses number sentences or formulas to represent a value |

Driving Off Into the Sunset
Grade Level: Upper

MATERIALS: Weather maps and road maps

ORGANIZATION: Cooperative groups of four

PROCEDURE: The purpose of this activity is to allow students to practice creating formulas using letters for unknown quantities.

The easier portion of this lesson involves use of road maps to create formulas to figure mileage. Have students select four different towns along a particular route and label the towns a, b, c, and d. From this point, have them create formulas which will tell the total mileage between a and d and other formulas such as the following:

1. How much farther (or shorter) is a to b than c to d?
2. If you got 20 MPG, how much gas would you use going from a to d and back to a?
3. If you can walk 4 miles in one hour, how long will it take to walk from a to c?

As a second portion of the lesson, have students use the information from the weather maps or weather reports to find how many hours and minutes between sunrise and sunset, between moonrise and moonset, or between the high and low temperatures for the day.

An example might be: If X = time of sunrise, and if Y = time of sunset, let Z = the amount of time between X and Y. A formula which will show how to find the amount of time from sunrise to sunset would be:

$$(12:00 - x) + (12:00 - y) = z$$

If x = 5:40, and y = 6:50, then

```
 12:00      12:00      5:10
 -5:40      -6:60     +6:20     z = 11:30 of sunlight
 -----      -----     -----
  6:20       5:10     11:30
```

Hands On, Inc
2121 Rebild Drive
Solvang, CA 93463

| 20 | Uses number sentences or formulas to represent a value |

Shoebox Algebra
Grade Level: Middle/Upper

MATERIALS: Shoeboxes, string, scissors

ORGANIZATION: Individually or groups of two

PROCEDURE: The variables in an equation represent those items which can change from situation to situation without changes in the overall relationship of the elements. This lesson uses string and a shoebox to teach this concept.

Ask the class how to find the perimeter of rectangle. They should explain that you measure four sides and add the figures. Some students may even explain that it can be done in terms of the formula $P = 2L + 2W$. Explain that students are going to find the perimeter of a rectangle in a unique way.

Divide the class into groups of two and give each team a shoebox, string, and scissors. Have your students cut a length of string equal to the length of one of the short sides of the box and another the length of one of the long sides. Ask them how many pieces of each string it would take to represent the perimeter of the shoebox. They should say that they need two of each, two short and two long. This could be written in a formula as $p = 2s + 2l$. This could be further simplified to $P = 2(S + L)$.

Next, explain that most shoeboxes are made so that the length is twice as long as the short side. This is not exactly true, but for the sake of this experiment we will generalize. With this knowledge, have your students rewrite the relationships between the long and short strings. An example might be as follows: since $l = 2s$, then $p = 2(s + l)$ might be written as $p = 2(s + 2s)$, or $p = 2s + 4s$, or $p = 6s$.

Have them measure one of the short pieces of string and see if this formula works as a generalization for all shoeboxes. Obviously this formula is not universal for finding all perimeters, but it will show how a formula can be created to solve for a constant relationship.

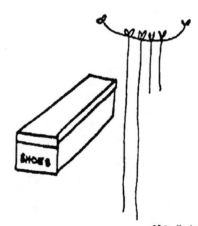

Hands On, Inc
2121 Rebild Drive
Solvang, CA 93463

Algebra

20	Uses number sentences or formulas to represent a value

You Can Even Add Odd!
Grade Level: Middle/Upper

MATERIALS: Tagboard, markers and string or yarn.

ORGANIZATION: Divide the class into teams of two

PROCEDURE: Before the lesson, the teacher should prepare tagboard cards (9' x 12") by writing operational signs on each one (+, −, x, =, and −). Students should have a quantity of blank tagboard signs of at least 8" x 5".

Divide the class into two teams and give each team a set of the operational signs. On the chalkboard the teacher should write, " e = even numbers, o = odd numbers, and n = number." The list of equations (see below) should be written on the chalkboard. The teacher should begin by saying "n = 6" and pointing to one of the equations below (o = o).

Team A will write three numbers on their tagboard and use the operational signs (i.e. 3 + 3 = 6 This team will be the first to present their equation in front of the class. If team A's equation is correct they receive one point. All team members should take turns writing and displaying.

Using the same number (6) the teacher points to a different equation on the chart. Team B writes numbers on their tagboard for this equation. A point is given for a correct response. Keep taking turns until all the equations have been illustrated for that number. The teacher or a student should then call out another number and the activity continues.

Students will find that certain combinations are not possible; for example, two even numbers cannot add, subtract, multiply into an odd number. Let students discover these relationship and have them explain why this occurs.

o + o	o − o	o X o	o − o
o + e	o − e	o X e	o − e
e + o	e − o	e X o	e − o
e + e	e − e	e X e	e − e

Hands On, Inc
2121 Rebild Drive
Solvang, CA 93463

Algebra

| 20 | Uses number sentences or formulas to represent a value |

Name That Tune
Grade Level: Upper

MATERIALS: A piano (if available), song bells, prepared music sheets (see below)

ORGANIZATION: Cooperative groups of four

PROCEDURE: This is a fun lesson especially for teachers of classes that are musically inclined. What students will be doing is using equations to identify particular notes on a piano or on song bells and they will have to put the notes together in order to identify the song.

We will use the example Twinkle, Twinkle, Little Star, and we will call middle "C" one, "D" two, "E" three, "F" four, etc. (as shown in the chart), Twinkle twinkle Little Star would be played by numbers one, one, four, four, five, five, four.

Give each student a copy of the song sheet (equation sheet) and show where they will be solving for the unknowns. Note one will be x, note four will be y, and note five will be z. Once the students have solved the equation, give each group the opportunity to play the song on the piano and try to name it. Once you have gone through the four songs with the students and your class understands the concept you may want them to create their own puzzles for other songs.

This concept of solving for the unknown becomes a game to the students and locks in the concept of a variable.

As an extension, with a class that is comfortable with working with positive and negative numbers, you might have students create equations which go below middle c. B could be negative one; A, negative two, etc.

A song sheet for <u>Twinkle, Twinkle Little Star</u>

$1(7+6) - 12 = n$
$1(3-2) = n$
$5(7+3) - 2(11+12) = n$
$(8-5) - (9-8) = n$
$4(5 \times 5) - 20 = m$
$(3+7) - (1 \times 5) = n$

Algebra

| 21 | Uses various methods to solve for variables in simple equations |

Equatiocard
Grade Level: Middle/Upper

MATERIALS: A deck of cards for each team of two students, gameboards (Appendix H)

ORGANIZATION: Teams of two students

PROCEDURE: The object of this card game is for students to get rid of all of their cards by creating simple equations.

Begin by dividing the class into teams of two students and give each team a deck of cards. Deal ten cards to each player. Let the players decide who will play first and begin the game with player 1 placing two cards on the gameboard and making a statement such as, "two + X = five."

Player two can then play a card if he/she has a three. If player two can "solve" the equation (has a three), then player two gets to place the next two cards down and make the next statement; if player two cannot solve the equation (no three in hand), player 1 can play a three. If player 1 has no three, then he/she must pick up the original two cards and start over again. Whoever solves the equation gets to make the next statement. Students will learn that there is strategy in using the = versus the < or > symbol.

Jack = 11, Queen = 12, and King = 13. The first several plays of a sample game is shown below.

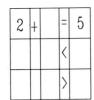
Player 1 plays

If player 2 has a 3, he plays and

then plays the next set

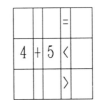
If player 2 has no 3, then player 1 can play a 3

If neither player has a 3, then player 1 picks up his cards and plays another set

If player 2 has a 10, J, Q, or K, he can play and then lay down the next set

Hands On, Inc
2121 Rebild Drive
Solvang, CA 93463

Algebra

| 21 | Uses various methods to solve for variables in simple equations |

Does Your Body Count?
Grade Level: Middle/Upper

MATERIALS: Large (12 X 12) pieces of tagboard and felt pens

ORGANIZATION: A whole class activity

PROCEDURE: This lesson shows students how to build examples of equations, to substitute concrete items for letters, and how to manipulate and solve equations.

Prior to the lesson, the teacher needs to prepare tagboard signs with one operation symbol on each (addition, subtraction, multiplication, and division). Signs for equal, greater than, less than, not equal, and parentheses should also be made.

At first, this activity should be teacher directed. Assign a student to hold a plus sign and another student to hold an equal sign and have them stand in front of the class. Tell two students to stand on the left side of the student with the plus sign and have three students stand on the right side of the plus sign student. Ask the class to send up enough students to stand on the right side of the equal sign student to complete the equation correctly. This is a very simple example.

As students understand the activity, more difficult examples can be set up. For example, have a division, multiplication, and equal sign diplayed and ask two students to stand to the left of the division sign. Remind students that the multiplication operation must be done first and that the product of the multiplication must divide evenly into two (you can't display 1/2 of a person).

The next step is to use greater than and less than signs so that students must find solution sets.

This can be played as a game with teams competing to form equations and asking the other team to complete them, or just as a fun activity to reinforce the concept of solving an equation.

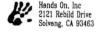

Algebra

| 21 | Uses various methods to solve for variables in simple equations |

Finding a Weigh
Grade Level: Middle/Upper

MATERIALS: Balance scale, envelopes, tagboard slips with numerals/variables/symbols

ORGANIZATION: Individually

PROCEDURE: Begin by demonstrating on a balance scale that the only way to keep in balance is by adding or subtracting equally to both sides. Point out that the fulcrum point acts as an equal sign between the two scale sides.

Give each student an envelope with small tagboard slips and have them write the numerals 1 - 12 on the slips. They should make two or three sets of these numbers. Also have them make several cards with +, -, x, - and =. Blank cards can be used as the variables in the equations.

On the board, write $3a + 2 = 11$ and have students place cards in front of them to mirror this. Explain that since there is an equal sign, the amount on each side of the sign must be the same. Since they do not know the value of the variable "a," they can do nothing with that part of the equation. They can, however, do something to the 2, they can remove it; BUT to keep the scale in balance, they must remove 2 from the other side as well. Have them display this equation $3a + 2 - 2 = 11 - 2$.

Point out the +2 and -2 will cancel one another and 11 - 2 equals 9. The students should be adjusting their tagboard slips throughout to mirror this procedure. The new equation will be $3a = 9$. To solve this, the student must have knowledge of inverse operations, but basically the process is the same. 3a must be divided by 3, and 9 must be divided by 3 to keep the scale "in balance." This final equation would be $3a - 3 = 9 - 3$.

Divide the class into teams of two, three, or four and have them go through this process with other equations, explaining each step as they go. Each time your students solve for a variable, remind them that they must always keep the scale in balance.

Hands On, Inc
2121 Rebild Drive
Solvang, CA 93463

| 21 | Uses various methods to solve for variables in simple equations |

A Good Sense of Spell
Grade Level: Middle/Upper

MATERIALS: No special materials are required

ORGANIZATION: Whole class activity

PROCEDURE: This activity uses the spelling rules of adding a suffix to words ending in "e" to explain the concept of solving for variables in an equation.

Begin by writing the following set of words on the board:

peace	ing
file	ment
blame	ful
state	er

Ask students to create a set of words combining the root in column one with the suffixes in column two. Ask one of the students to come forward and write a few responses on the board.

Ask them to identify similarities in the words in column one, eventually eliciting that all end in "e" preceded by a consonant. Explain that this can be written as "c + v." Next, point out that two of the suffixes begin with the pattern v + c, and two begin with c + v and ask if there is any pattern to the way in which the words they have written are spelled. The ultimate goal being for students to create the pattern:

$(c + v) + (v + c)$ = incorrect spelling   e.g.   file + er = fileer
$(c + v) + (c + v)$ = correct spelling   e.g.   peace + ful = peaceful

Students should attempt to state this relationship algebraically, thus creating a "spelling formula" for silent "e" -- $(c+v) = v = e$ OR $(c + v) + c = e$. Have students try this same procedure with other spelling rules such as "i before e" or making plurals of words ending in y.

Hands On, Inc
2121 Rebild Drive
Solvang, CA 93463

Algebra

22	Uses various methods to solve for variables in complex equations

Call for an Operation
Grade Level: Upper

MATERIALS: No special materials are necessary

ORGANIZATION: Individually

PROCEDURE: In this exercise, students will be using their phone numbers as a source of numbers to create equations and then solve for a given variable.

Explain to students that numbers can be grouped or structured in various ways to solve for a variable. This lesson is going work at this in a somewhat "back to front" format in that they will be given the numbers and will have to decide the operations and order of computations.

Tell students that their task is, given the order of numbers in their phone numbers, to construct an equation which will result in the variable a (in the first case, a = 0). How can they use operations, parentheses, or brackets to write their phone numbers so that it will equal a? We will use the Hands On Phone number 688-0089 as a sample.

$$[(6 \times 8 - 8) + 0] \times 0(8 + 9) = a$$

Once students understand the concept, have them solve for b (b = 1) or c (c = 2) etc. Students will find this activity challenging, but with creativity they will be able to create solutions for almost any variable < 10.

Hands On, Inc
2121 Rebild Drive
Solvang, CA 93463

Algebra

| 22 | Uses various methods to solve for variables in complex equations |

An Eye for an Eye and a Bean for a Bean
Grade Level: Upper

MATERIALS: An assortment of beans (four types), enough for sixty beans for each group

ORGANIZATION: Teams of two students

PROCEDURE: In this activity, students will use the method of subtracting equally from both sides of an equation to arrive upon the value of a variable. Students should be concerned with the PROCESS rather than the PRODUCT in this activity.

Begin by giving out a pile of sixty assorted beans to each group. The students should separate the beans into two random, but equal piles (30 beans each).

Next, count the number of each type of bean in each pile and record the results on a piece of paper. Tell them that the two piles are equal in value and their job is to find out the value of one type of bean in relation to the other beans. As an example, we'll use the following configuration:

    PILE A: 16r (red beans), 7k (kidney), 3p (pinto), 4n (navy)
    PILE B: 9r, 9k, 8p, 4n

Since $A = B$, then $16r + 7k + 3p + 4n = 9r + 9k + 8p + 4n$

The next step is for students to simplify the equation by subtracting (eliminating) like beans from each side of the equation; however they should avoid negative numbers unless they already understand the concept.

We will begin by removing 3p from both sides: $16r + 7k + 4n = 9r + 9k + 5p + 4n$.
Now remove 7k from both sides: $16r + 4n = 9r + 2k + 5p + 4n$
Next, remove 4n from both sides: $16r = 9r + 2k + 5p$
Finally, remove 9r from both sides: $7r = 2k + 5p$ THEREFORE, $r = \dfrac{2k + 5p}{7r}$

Algebra

| 22 | Uses various methods to solve for variables in complex equations |

Do Unto Others...
Grade Level: Upper

MATERIALS: Balance scale, weights or objects that can be used as weights (nuts, bolts, etc. -- at least 50 weights per group)

ORGANIZATION: Teams of two

PROCEDURE: This is an extended lesson which may take two or more days to complete. It incorporates the the concept of equality in an equation. The lesson has four parts: addition, subtraction, multiplication, and division.

Begin with a whole class demonstration in which the teacher places five weights in each of the pans thus balancing the scale. Show students what you are doing and ask them to generate an equation depicting this setup: 5 = 5. Now add two weights to one pan and point out that the scale has lost its balance. This will not be too surprising to the students, but elicit from them the way in which to rebalance the scale (add two to the other pan). Have them write this equation: 5 + 2 = 5 + 2.

Point out that as elementary as this process may seem, it is the underlying concept in solving equations. Students must learn to always add to one side the same amount they have added to another.

From this point, ask students about the subtraction of weights in the pans. Remove four weights from one pan and ask how they might balance the pans (remove four from the other side). The next step will take this basic concept to a working level.

5 = 5

5 + 2 = 5 + 2

7 = 7

7 - 4 = 7 - 4

Hands On, Inc
2121 Rebild Drive
Solvang, CA 93463

Algebra

Ask students how they might use the scale to solve an equation such as: $5 + n = 8$. They will undoubtedly be able to determine that $n = 3$, but the step by step process is the purpose of the lesson.

Explain that in an equation, the goal is to isolate the unknown on one side of the equal sign. To do this, they will do the inverse operation of addition (subtraction) and will need to subtract from both sides of the scale.

$5 + n = 8$

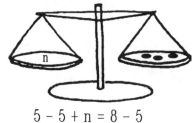
$5 - 5 + n = 8 - 5$

In order to balance the scale, the students will need to add three weights to replace n; therefore, $n = 3$.

Give students several other examples to work on in addition before proceeding.

In subtraction, students will once again use the inverse operation (addition) to solve for an unknown. Begin with the equation: $n - 5 = 2$. Students will again be able to identify that $n = 3$, but have them demonstrate this on the balance scale.

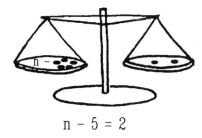
$n - 5 = 2$

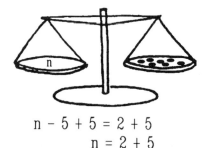
$n - 5 + 5 = 2 + 5$
$n = 2 + 5$

Explain that $-5 + 5$ cancel one another and therefore, seven weights must be added to balance the scale, so $n = 7$. Next, have students try to solve for $5 - n = 3$. This organization cannot be demonstrated on the balance scale because the solution becomes $-n = -2$ ($n = 2$). Although the balance scale cannot depict this relationship, students can see the need to treat both sides of an equation equally.

The concept of multiplying and dividing equally to both sides of an equation can be taught using the balance scale but solving for unknowns as in the above examples is best taught through other methods shown in this book.

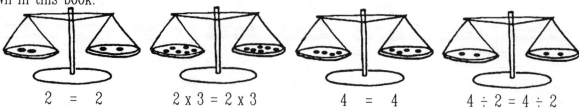

$2 = 2 \qquad 2 \times 3 = 2 \times 3 \qquad 4 = 4 \qquad 4 \div 2 = 4 \div 2$

Hands On, Inc
2121 Rebild Drive
Solvang, CA 93463

| 23 | Uses equations to solve problems with decimals or fractions |

Have It Your Way
Grade level: Upper

MATERIALS: A Burger King menu with prices, order blanks like those used by waitpersons

ORGANIZATION: Cooperative groups of four

PROCEDURE: In this activity, students will imagine that they are Burger King order takers and they will use the information on orders to create equations.

Begin by dividing the class into cooperative groups of four and give each student an order blank. One by one, each student should allow each other person in the group to place an order, including his own, so that each student has orders form four people. Given this information, each group member will write an equation to determine the total price of lunch by placing a value (menu price) on each unknown.

As students complete their computations, have them compare their totals with one another. A sample group might create the following list:

Student	Order	Equation for order
Yvette	1 Whopper, fries, small beverage	$W + F + S$ = Yvette
Slim	2 Burger Deluxe, fries, apple pie, large beverage	$2B + F + P + L$ = Slim
Raul	1 chicken sandwich, fries, large beverage	$C + F + L$ = Raul
Mimi	2 Whoppers, 2 fries, apple pie, diet beverage	$2W + 2F + P + M$ = Mimi

Total group equation: $W + F + S + 2B + F + P + L + C + F + L + 2W + 2F + P + D$ = Price
Simplified group equation: $3W + 5F + S + 2B + 2P + 2L + C + D$
Equation with values for unknowns: $3(1.89) + 5(.79) + .69 + 2(1.45) + 2(.89) + 2(.99) + 1.67 + .79$ = Price

The final step will be to combine the entire class into one order and to figure tax on the total cost.

The benefit of this type of activity is that students can see that algebra is something they actually use every day. Discuss this with the class as you begin the process of compiling each group's equations. As an extension, you might have students relate the price of an item (i.e. fries) as a fractional part of another (i.e. burger).

Hands On, Inc
2121 Rebild Drive
Solvang, CA 93463

Algebra

| 23 | Uses equations to solve problems with decimals or fractions |

All Mixed Up
Grade Level: Upper

MATERIALS: Assorted nuts, trail mix items, or m & m's (bought separately, not as a mix), price lists of each item separately and by weight or amount.

ORGANIZATION: Cooperative groups of three or four

PROCEDURE: This activity should be demonstrated and modeled by the teacher for the whole class before they work in small groups. The teacher should pose the problem that students are to make a mixture of items that will cost 75 cents.

Using the nuts as an example, the mixture must include different kinds: peanuts, cashews, almonds, etc. Each type of nut is a different price per pound. The students need to find the cost and weight of each individual type of nut to find how many of each type can be put in the mixture to make the price 75 cents. For example, few cashews will be put in the mixture because they are expensive. More peanuts will be included because they are cheap. A sample problems shown below.

peanuts are $1.75 per pound and there are 215 peanuts per pound
cashews are $4.75 per pound and the are 112 cashews per pound
almonds are $3.49 per pound and there are 139 almonds per pound

Given this information, have students write equations for figuring the cost of one nut of each type.
p = 1.75 ÷ 215   c = $4.75 ÷ 112   a = $3.49 ÷ 139
p = $.008   c = $.04   a = $.025

At this point, students will need to use trial and error to create a combination which totals $.75. Once they have created the combination, have them write this as an equation. Given the sample items above one possible formula would be: 8c + 12a + 16p = $.748 = $.75.

Once students understand, they can work the other way around and give them a trail mix and have them figure the individual cost of each ingredient.

Hands On, Inc
2121 Rebild Drive
Solvang, CA 93463

Algebra

| 23 | Uses equations to solve problems with decimals or fractions |

Tricky Taxes
Grade Level: Upper

MATERIALS: Catalogs, scissors, chart paper, glue

ORGANIZATION: Individually

PROCEDURE: This lesson gives students practice in using tax rates from various states to compute total cost of items.

Give each student a magazine or catalog to select three items for purchase and ask them to write a formula which will allow them to figure 5% sales tax on each of these items. A sample equation using i for item purchased, t for tax rate, and c for total cost might be: $i + (i \times t) = c$.

Have them do the computation for tax rates of 5%, 6% and 7%. Once this complete, have students make a collage or pictorial representation of comparative costs of these three rates. Imagining that Arizona has a tax rate of 5%, California has a tax rate of 6%, and New Mexico a rate of 7%, a sample picture of this might be:

Algebra

| 23 | Uses equations to solve problems with decimals or fractions |

Body Weight
Grade Level: Upper

MATERIALS: One or more bathroom scales

ORGANIZATION: Whole class or small group activity

PROCEDURE: Your class will have fun with this one in that they will be weighing themselves and creating equations which will estimate the weight of different extremities of the body.

Select a brave volunteer from the class to come forward and be the "guinea pig." Have the student recline on a table and place his right arm over the bathroom scale. Have him apply no pressure and write down the weight of this arm. Repeat with the left arm.

Have students estimate the percentage of total body weight contained in the arms and repeat this process for the right and left legs. Once again, have students estimate the total percentage of body weight.

Next, have the student place his head on the scale (laying on his back and resting it on the scale) and write down this weight.

Given this information, and the total weight of the volunteer, have the students create an equation which represents the percentage of weight distributed to each extremity and the torso.

Divide the class into groups of four to six students and let them repeat the process within the group to see if the formula is accurate. You will need to be sensitive to the unwillingness of some students to be weighed; nevertheless, all students can participate in the mathematical portion of the lesson.

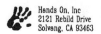

Algebra

| 24 | Uses linear equations to represent relationships |

A Penny a Pound
Grade Level: Upper

MATERIALS: A stump or large piece of wood, an assortment of nails starting with 6 penny and going all the way up to 16 penny nails

ORGANIZATION: Whole class activity

PROCEDURE: In this activity the students will try to discover whether hammering nails, the length of the nails, and the number of hits will result in a linear equation. The class will find that for some students it will, while for others it will not. This concept of "what a linear equation is not" is as important as an explanation of a real linear equation.

Select a volunteer from the crowd to come forward and pound a six penny nail into a stump. Have the other students estimate and then count how many hits it will take. Have the same student use the number eight nail and drive it, then a ten and a twelve.

At this point in time, see if there is a pattern which has emerged (see chart below). Plot this information on an overhead and see if it does indeed create a line, or a linear motion. You will probably find that for most students the graph is linear for a time but then will change. This is a good opportunity to talk about variables in terms of accuracy, strength of hit, diameter of the nail, the ratio of hits to misses, margin of error, etc.

Nail Size	Number of Hits	Number of Misses
6 penny	8	1
8 penny	16	1
10 penny	26	4
12 penny	44	9
14 penny	68	12

Ask students how this same exercise could be done to create a linear equation. For example, if they had a machine that was accurate and hit with the same force, would there indeed be a linear equation? Have other students try their hands at hammering nails.

Hands On, Inc
2121 Rebild Drive
Solvang, CA 93463

| 24 | Uses linear equations to represent relationships |

Linear Circles?
Grade Level: Upper

MATERIALS: A collection of various sized wheels such as a bike tire, wheel barrow tire, wagon wheel, toy car wheel, etc. (four or five wheels is enough), graph paper, measuring tape, and yard or meter stick

ORGANIZATION: Whole class activity or in cooperative groups

PROCEDURE: In this activity students will the ratio of radius to number of revolutions to create a linear equation.

Begin by holding up the bike tire and asking students to figure the circumference. Let them measure the radius of the tire and compute an answer using the formula c = 2r x pi.

Once students have agreed upon the correct circumference, have them estimate the number of times the wheel would have to revolve in traveling from one end of the classroom to the other. Let them actually roll the wheel to check their estimates. Have them graph this information as shown below.

Discuss the question as to whether there is a set ratio between the radius of a wheel and the number of revolutions to roll across the room. This is the task the students will be solving.

Give each group a wheel and have them compute the circumference and number of revolutions, each time graphing the result. Once each group has three plotted points on their graph (bicycle plus two more), have them arrange the dots and estimate the number of revolutions for any unmeasured wheels.

What interesting line do they create in their measurement?

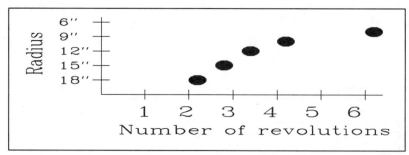

140

Algebra

| 24 | Uses linear equations to represent relationships |

Getting the Drop on Linear Equations
Grade Level: Upper

MATERIALS: Five balls that bounce (tennis, ping pong, etc.), tape measure/stick, and grid paper

ORGANIZATION: Pairs or groups of four

PROCEDURE: Explain to your students that a linear equation is one in which the variable, when plotted on a graph, will result in a straight line. In this activity, students will do an experiment in which the results can be represented with a linear equation.

Tell students that each group is to take a ball and drop it from a certain height and measure its bounce. For the sake of ease, and to lessen human error, have them do this against a wall and drop in specific increments (inches/centimeters, etc.). Each team should use a different type of ball and comparisons should be made at the end of the lesson with students creating a graph similar to that shown.

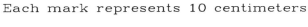

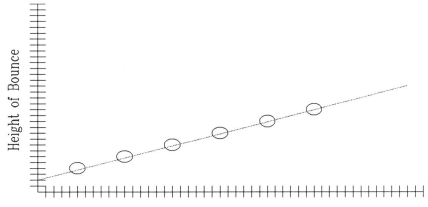

Algebra

| 24 | Uses linear equations to represent relationships |

I've Gotta Split!
Grade Level: Upper

MATERIALS: Split peas, construction paper, glue, maps of the United States

ORGANIZATION: Cooperative groups of three or four

PROCEDURE: In this activity, students will create a table which describes a linear equation and will then write equations which describe the table.

Begin by dividing the class into groups of three or four. Give each group a map of the United States, a supply of split peas, and a sheet of construction paper. Next, assign each group a different "home base" city. In this example, we'll use Las Vegas.

Have each group imagine that they are postal employees who assign postage rates to packages sent thoroughout the United States. The rates are determined as follows: The rate for a 1 pound (1 kilo) package is 1 split pea per 250 miles (kilometers). From this information, students are to complete the table shown below.

In completing the table, the students should glue the peas to the paper, thus providing a concrete example from which to write their equations.

As an extension, you may wish to have groups double the cost per 250 miles (1 pea, 2 peas, 4 peas, etc.) or triple the cost (1 pea, 3 peas, 9 peas, etc.).

City in mileage zone	# of Miles	Cost (in peas)	Equation
Los Angeles	280	2 peas	p + p = 500
San Francisco	440	2 peas	p + p = 500
New York	2655	11 peas	11p = 2750

Hands On, Inc
2121 Rebild Drive
Solvang, CA 93463

Algebra

| 25 | Creates equations employing parenthesis and brackets |

Greater Than a Book
Grade Level: Upper

MATERIALS: Book shelves with books and bookends.

ORGANIZATION: Groups of two students, or large group activity

PROCEDURE: This is a basic lesson giving students a concrete example of the purpose of parenthesis and brackets. It will also provide review in the Dewey Decimal system of numbering in the library.

Take the students to the non-fiction section of the library and have each student choose a shelf of a bookcase. This shelf will generate some numbers that can be used to write an equation, and the bookends that hold a group of books represent parentheses or brackets.

Have students scan the shelf, noting the numbers on the book bindings. In the 100 -199 section, for example, students will find books numbered 100.2, 100.3, 110.3, etc. In some cases, they will find several books with the same number. Their task is to write an "equation" for the library shelf with each like numbered book - with the same tenth or hundredth in parenthesis; books with the same three digit whole number in brackets, and the entire 100 - 199 section within braces. A simple example is shown below.

$$\{[100.1 + (100.2 + 100.2) + 100.6] + [(110.2 + 110.2 + 110.2) + 113.4 + 114.3]\}$$

At this point, you may want to have students simplify their equations, i.e. 2 x 100.2, and place them on sentence strips to display in class.

This information can lead to a discussion of library organization and why there are few books in certain non-fiction numbers and many books in others.

Hands On, Inc
2121 Rebild Drive
Solvang, CA 93463

| 25 | Creates equations employing parenthesis and brackets |

A Well Balanced Diet
Grade Level: Upper

MATERIALS: 3 x 5 cards

ORGANIZATION: Individually or in pairs

PROCEDURE: In addition to performing operations in the order of MDAS (multiplication, division, addition, and subtraction), certain symbols determine the order of computation in equations. In this activity students will learn how to put the elements in an equation in proper order.

The order of operations in symbols is: parentheses first and brackets second. It is important to add that within the parentheses or brackets the order of MDAS should be followed.

Distribute three 3 x 5 cards to each student and have them list their favorite breakfast on card one, lunch on card two, and dinner on card three. Limit the number of items to three on each card. Pass the cards to a partner who should assign a variable (see below) to each breakfast item.

eggs, bacon, and potatoes on the breakfast card = $e + b + p$
king kong burger, fries, and chips on the lunch card = $k + f + c$
lasagna, salad, and ice cream for dinner = $l + s + i$

Randomly lay the three cards out and write the equation presented. If the card order is lunch, breakfast, dinner, the equation becomes: $k + f + c + e + b + p + l + s + i$. In order to get them in time order use parentheses to indicate which comes first: $k + f + c + (e + b + p) + l + s + i$. Next, use brackets to indicate which comes second: $[k + f + c + (e + b + p)] + l + s + i$.

The next step is to have equations combined between the team members. For example, if both students like eggs and bacon, but student two likes toast, the breakfast equation would be $(2e + 2b + p + t)$.

Hands On, Inc
2121 Rebild Drive
Solvang, CA 93463

Algebra

| 25 | Creates equations employing parenthesis and brackets |

Don't Lose Your Cookies
Grade Level: Upper

MATERIALS: Copies of cookie sheets (Appendix I), scissors, bag (to draw from)

ORGANIZATION: Individually

PROCEDURE: In this activity students will sharpen their ability to recognize which operation to complete first, a skill which is made simple by the use of parentheses and brackets.

Have students cut the cookie cut-outs from the appendix and place them into the bag. Next, have each student pick a cut-out at random. Ask them how many individual cookies each student has. Each cookie package has a different number of cookies but all are in two rows, this can be represented by 2 x c. Now ask the class how many students have the same type of cookie package. This is expressed by using parentheses around the original statement multiplied by the number of students: for example, if five students have the same type of cookies it would be written as 5 x (2 x c).

The next step is to ask students how much they think their bag of cookies would weigh if it were real. Take estimates as to the weight of a normal pack of cookies found in the grocery store and select a weight to use for the next step in the formula (or you could use an average of all the estimated weights). So far our formula shows us what the total number of cookies in the class is. In order for us to find the weight of all these cookies, we would multiply our formula by the estimated weight. This step in the formula needs to be separate from the others and should be performed last. To indicate this situation algebraically we use brackets. If the estimated weight is 20 ounces, then it would be written 20 [5 (2 x c)] = total weight in ounces (these weights can easily be exchanged for metric).

As an extension, students can do research at the market to check their estimations. They can also find out the number of calories for each individual cookie by using a similar process (you can also measure sugar content, oil/fat content, salt content, etc. of any product).

Hands On, Inc
2121 Rebild Drive
Solvang, CA 93463

Algebra

| 25 | Creates equations employing parenthesis and brackets |

Anything but Chili
Grade Level: Upper

MATERIALS: Simple recipes from home for each student (four or five steps--you may need to bring spares from home), scissors, food magazines or newpaper sections

ORGANIZATION: Individually

PROCEDURE: Normally in math, students perform operations in the order of multiplication, division, addition, then subtraction. In many situations, however, certain parts are dealt with as if they are separate. This is expressed in algebra by the use of parentheses, brackets, and then braces (in that order). In this activity students will use recipes from home to demonstrate this concept.

Have students cut out their recipes and glue them to a piece of paper to work on. Next they should assign each item in the recipe a variable (see the example below). They should then group items using parentheses to represent which ones go together and which ones are dealt with separately. Groups which are processed differently should be separated by brackets. The final step of baking, blending, or mixing should be separated by braces.

Chili Inferno

1 lb. raw italian sausage (assigned the variable I), 1 large Bermuda onion (b), 4 jalepenos (j), 2 large carrots (c), 1 lb. cooked pinto beans (p), 2T. chili powder (r), 1T. cayenne (x), 6T. paprika (h), 3T. oregano (o), 3tsp. salt (s), 3tsp. ground black pepper (h), 10 oz tomato paste (t), 14 oz tomato sauce (s), 16 oz stewed tomatoes (l), and 2 cloves of garlic (g). Chop all vegetables.

Cook sausage in crock-pot and drain off fat (f). Pour the tomato sauce into crock-pot, add the tomato paste and stewed tomatoes and let simmer over a low fire. Next, add pinto beans. When the sauce begins to bubble, add all spices and stir and let simmer thirty minutes. Add chopped vegetables. Let simmer until beans and vegetables are tender. Let simmer for an hour or two then serve.

The formula is: Chili = $\{[(i - f) + (t + s + l)] + p + (j + r + x + o + h + s + g)\} + (b + c)$

Have students cut out pictures from magazines or newspapers to represent the ingredients of their recipes (some may need to be drawn). Then have them glue these onto a piece of paper in the from shown in their formula.

Hands On, Inc
2121 Rebild Drive
Solvang, CA 93463

Algebra

| 26 | Creates equations and number sentences including symbols: <, >, = |

On a Roll
Grade Level: Upper

MATERIALS: Four dice for each group of two students, cards for writing greater than, less than, equal to, +. −. X, and ÷ symbols

ORGANIZATION: Teams of two students

PROCEDURE: This activity gives students practice in creating equations from random numbers. Since the lesson is somewhat difficult, you may wish to divide the class into groups of four with two students working together on each team.

Give each set of players four dice and several cards. They should make two cards for each of the symbols described in the materials. Player 1 begins by rolling the four dice. The object is to use the numbers on the dice to create a number sentence. If the student can create a sentence which uses an equal sign (and it is correct), player 1 receives 5 points. If the student creates a sentence using < or >, then student 1 receives 3 points. Once student 1 has finished the sentence, player 2 repeats the process. The first student to receive 25 points is the winner.

To create these sentences, the students can use each die individually, or can place them together in a place value format (eg., 2 and 3 could become 23 or 32).

Some sample rolls and solutions are presented below.

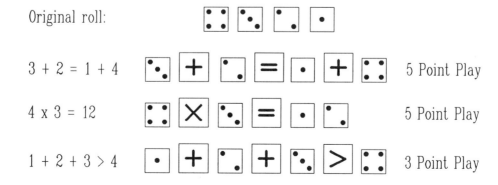

Hands On, Inc
2121 Rebild Drive
Solvang, CA 93463

Algebra

| 26 | Creates equations and number sentences including symbols: <, >, = |

Why?
Grade Level: Upper

MATERIALS: No special materials are needed

ORGANIZATION: A whole class activity

PROCEDURE: Middle school students are in the unique situation of never quite knowing how their friends will treat them each day when they arrive at school. Some days begin wonderfully and end in chaos, while some begin horribly and end in ecstasy. Since you teach middle school aged students, you already know this. In this activity, students will use these peer situations to create a lesson in which they work with >, <, and = symbols.

Begin by asking students if they have ever come to school expecting to have a good day, only to find that several (or one) of their best friends aren't speaking to them. From this introduction, have your class generate a list of reasons as to why this situation might have occurred. Try to come up with at least ten possible reasons, writing them on the board as they are suggested.

Ask students if there are some reasons that are more realistic than others; give them some time to discuss a possible "heirarchy" within the list. Tell students that there is a way to label this heirarchy in mathematical terms using the equal, greater than, and less than symbols. At this point, label the reasons A-J and have each student write a heirarchical list using these symbols. An abbreviated sample set is shown below.

Reasons for being ignored by your friends:
A. Spoke to someone else's boy/girl-friend.
B. Did not call the previous night.
C. Received a better grade on a test.
D. Told a secret to another friend.
E. Did not share a secret with another friend.
F. Wrote all over someone else's notebook.
G. Didn't invite a particular person to a party.

Hierarchical rankings:

$A > D = E > G > B > F = C$

Hands On, Inc
2121 Rebild Drive
Solvang, CA 93463

Algebra

| 27 | Creates equations using variables |

Wheeler Dealer
Grade Level: Upper

MATERIALS: A bike tire is a nice visual

ORGANIZATION: Whole class activity or in cooperative groups of four

PROCEDURE: Begin by holding up a picture of a bike tire (or display the real thing). Ask students to estimate the distance around the circumference of the tire. Once you have recorded their estimates show them the formula for computing the circumference $C = 2r \times pi$.

Let them measure the radius of the tire and compute an answer. This may also spark some discussion as to where the term "pi" comes from.

Once students have agreed upon an answer, have them estimate the number of times the wheel would have to revolve in travelling from one end of the classroom to the other. Let them actually roll the wheel to check their estimations.

Next, ask them if they could create a formula which would solve this problem for them. At this point, divide the class into cooperative groups and have each group find a circle in the room. As a group, they should go through the same procedures as were done in the bike tire example. Their ultimate goal is to write a formula which will tell the number of revolutions by any circle shape in order to travel one mile.

Give students time to struggle with finding a solution. The task is difficult but in working with the circumference formula and in creating a formula for revolutions in a mile, students get practice in understanding how equations work.

> one mile = 5280 ft.
> 5280' x 12" = 63360"
>
> 63360"/r" x 3.14 =

Hands On, Inc
2121 Rebild Drive
Solvang, CA 93463

Algebra

| 27 | Creates equations using variables |

Name Games
Grade Level: Upper

MATERIALS: No special materials are required

ORGANIZATION: Cooperative groups of four

PROCEDURE: Explain to your students that variables are combined in various ways to represent a "whole" or result. In this activity they will be developing their ability to create equations.

Begin by putting all the letters of your first name on the board. Represent any letters that are repeated by a multiple ($a + l + l + a + n$ = Allan, would be written as $2a + 2l + n$ = Allan). Tell students that this is known as simplifying an equation. Have the students do the same thing with their own names.

After they have become familiar with the idea, have each group combine all their names and then simplify. You can expand this activity to the whole class.

Another extension could be to assign a value to each letter and to get totals for different names or classes. You might also randomly assign positive and negative numbers to the variables.

Allan + Juan + Marisol + Annette

$5a + 3l + 4n + j + u + m + r + i + s + o + l + 2e + 2t$

Algebra

27	Creates equations using variables

Rebus Variables
Grade Level: Upper

MATERIALS: Magazines, scissors, and glue

ORGANIZATION: Individually or in pairs

PROCEDURE: Explain to your class that in algebra letters are used to represent elements in number sentences, also called equations, which may change in different situations. The equation itself represents a relationship between these elements.

Ask a student to find a picture of something in a magazine (we will use a car as an example) and have them identify the component parts of a car (wheels, doors, windows, lights). Assign each of these elements a letter to identify it as a variable. An example might be: $w + d + g + l = c$ (wheels + doors + glass + lights = car). Point out that a car company could combine any variety of different doors with wheels, etc. to create their cars. These elements vary greatly but are in a constant relationship which results in a car.

The job for your students is to find a picture of an item which has at least three component parts. They should identify each variable with a letter and create an equation for the picture. Their next step is to find other photos of these items, cut them out, and create a "rebus" (equation) of the original item.

Using the car as an example, a student should find different pictures of wheels, glass, doors, and lights, and place them into an equation. The purpose is to show students that variables can change, but the end result of the equation remains the same.

Hands On, Inc
2121 Rebild Drive
Solvang, CA 93463

Algebra

27	Creates equations using variables

A Group "Outing"
Grade Level: Upper

MATERIALS: Various state road maps, copies of menus from several restaurants

ORGANIZATION: Cooperative groups of varied numbers (2 – 6 students)

PROCEDURE: Begin by dividing the class into groups of various sizes. Because students will be doing averaging, varied numbers will make the group reports more interesting.

Tell the students that they are going to plan a day long outing for their group. They will leave early in the morning and arrive home in the evening. They will eat breakfast, lunch, and dinner at restaurants along the way and will need to get gasoline (set the same gas price per gallon for each group) at various times depending upon the number of miles travelled.

The goal is for students to write a set of equations which can be used to figure the total cost of the day, the amount spent per person for food, the average cost per mile, and the total cost of the morning vs. the cost of the afternoon (after 12:00 P.M.).

Students should meet together to plan their excursion, writing down their menu selections and the cost of gas at each stop. They should then compile this information as a group. Allow them to figure and total the costs of each item mentioned above if they are having difficulty with writing the equations. Once students have arrived upon the totals, have them write the four equations. They should share this information with the class.

x = Gas stop #1
y = Gas stop #2
z = Gas stop #3
b = Breakfast
l = Lunch
d = Dinner
n = Number of miles driven

*Total cost for day: $(b + l + d) + (x + y + z) = t$
Amount per person for food: $(b + l + d) \div 4 = f$
Average cost per mile: $\{(x + y + z) \div \$1.20\} \div n = c$
Cost of morning: $t - (l + d + z) = m$

* There are numerous solutions, allow students an opportunity to discover this.

Hands On, Inc
2121 Rebild Drive
Solvang, CA 93463

Algebra

| 28 | Creates equations from word problems |

Word Problem Lotto
Grade Level: Upper

MATERIALS: Sets of word problems from old math books, small slips of paper for each student

ORGANIZATION: Cooperative groups of two, three, or four students

PROCEDURE: Students will be using sets of word problems from old math books to create puzzles for their classmates.

Divide the class into teams of two, three, or four students and give each group a set of three to five word problems and a stack of small cards. What the students will do is read each word problem and write one number or operation of the equation needed to solve the problems on a separate slip of paper. A sample might be:

1. Three business people were discussing their salaries over lunch. One mentioned that he made $45,000 per year, the second said, "I make $6,000 more than that," and the third said, "I make as much as both of you combined." What was the toal amount of salary received by the three people?

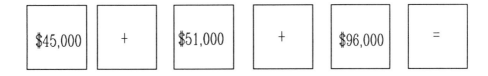

Once students have completed a set similar to this for each of their problems, they should mix all of the cards together for all problems and pass the cards and problem sheet to the next group to see if they can reassemble the cards in the correct order. Continue to circulate, giving each group an opportunity to solve several sets of problems.

Hands On, Inc
2121 Rebild Drive
Solvang, CA 93463

153

Algebra

| 28 | Creates equations from word problems |

Dear Abby
Grade Level: Upper

MATERIALS: No special materials are necessary

ORGANIZATION: Individually

PROCEDURE: Most students will be familiar with Dear Abby or Ann Landers columns, but if not, you might bring a few samples to read aloud.

In this activity, students will be writing to Dear Abby to explain a problem -- a math problem -- which is bothering them and will ask Abby for advice. The ultimate goal is to put all of the Dear Abby letters together into a "column" and let the students solve the problems (equations) presented in the letters.

Begin by reading a sample problem:

> Dear Abby,
>
> I have a problem about our school playground. There are several groups of students that like to play different games, but we don't ever seem to have enough room. The playground is 54 meters wide and 125 meters long. There is an area which is blacktop which is 24 meters by 50 meters. Our principal says that one third of the students must play on the blacktop and two-thirds must play on the grass. My problem is that if we have 240 students at school, how many square feet of grass area does each student have of his own? If you can help me, please write.
>
> Signed,
>
> Squashed

Have each student write a math problem in letter form and then distribute these to the class, having them create formulas and solutions for each letter.

Hands On, Inc
2121 Rebild Drive
Solvang, CA 93463

Algebra

| 28 | Creates equations from word problems |

Roamin' Around the Numerals
Grade Level: Upper

MATERIALS: Index cards with the Roman numerals I, V, X, L, C, D, and M

ORGANIZATION: Individually or in teams of two, three, or four

PROCEDURE: This lesson uses the Roman numeral as a word problem. Students must rewrite the numeral as an equation.

Begin by reviewing the value of each of the symbols in Roman numerals: I = 1, V = 5, X = 10, L = 50, C = 100, D = 500, M = 1000 and a line – over the top of any symbol, multiplies that value by 1000. Also review the term "arabic numeral" which is our numeration system of 1, 2, 3, etc.

Ask students to create a set of rules which govern the use of Roman numerals. They should include:

1. A letter which come before a letter of greater value means to subtract the first value from the second. i.e. IV means 5 – 1
2. A letter which comes before another letter which is of equal or greater value means to add the first value to the second. i.e. XX = 10 + 10 or XV = 10 + 5

Given these two rules, have students practice writing equations given some simple numerals such as:
    a) IX = 10 – 1   b) IVX = (10 – 5) – 1   c) XLC = (100 – 50) – 10

Once students understand the concept, give them a set of Roman numeral cards and have them randomly place several of them in a line. For this information, have them write an equation which solves for the value represented by the set of cards.

For example: CDXCIII = (500 – 100) + (100 + 10) + (100 + 3) = 493
                      MDCLXXV = 1000 + (500 + 100) + (50 + 10 + 10 + 5) = 1675
                      MV = 5000 – 1000 = 4000

Hands On, Inc
2121 Rebild Drive
Solvang, CA 93463

Algebra

| 28 | Creates equations from word problems |

Now That's a Tall Tale
Grade Level: Upper

MATERIALS: Several resources for reading tall tales about Paul Bunyon, Pecos Bill, Johnny Appleseed, Joe Magarac, John Henry, etc., access to encyclopedias

ORGANIZATION: Cooperative groups of four students

PROCEDURE: We usually think of tall tales as part of the literature curriculum, but in this activity, students are going to search for exaggeration in the tales and then try to prove, mathematically, why they are exaggerations.

Begin by reading a chapter of one tall tale to the students. Most tall tale literature will discuss one of the hero's escapades in each chapter of the book. Each team should work with only one chapter. For our example we'll use the story of Pecos Bill lassoing a tornado and riding it until tamed.

Ask students, as a whole group, to list exaggerations in this story. They might generate items such as, you can't lasso air -- the substance of a tornado, rope would never be strong enough to harness the power of a tornado, a person couldn't ride a tornado because it is too large to straddle, etc.

Now ask if any of these exaggerations can be proven false in a mathematical sense, i.e. can an equation be written which demonstrates the impossibility. For example:

A rope which Pecos Bill would have used would have had to have been made from a natural material since no synthetic materials were then available. Also, the rope used by any cowboy could not have exceed 1/2 inch in diameter because it would have been too heavy to throw for any distance. A rope of this size can subdue an animal up to 1000 pounds or 455 kilograms (World Book).

The speed of a spinning tornado is greater than 300 mph or 480 kph and they can easily shatter an entire house into pieces with this force. Therefore, Pecos Bill must be exaggerating because a rope with breaking point < 456kg could not possible contain a 480 kph wind.

Let students select a tall tale story to analyze within their group and share responses with the entire class.

Hands On, Inc
2121 Rebild Drive
Solvang, CA 93463

Algebra

| 29 | Creates equations which reflect number patterns |

Cubist Math
Grade Level: Upper

MATERIALS: Sugar cubes or other blocks

ORGANIZATION: Individually or in pairs

PROCEDURE: Explain to your students that a formula expresses a relationship or pattern. In this activity students will express a formula algebraically in an equation with variables to represent the elements which change or vary.

Begin by explaining that a cube is a geometric shape which not only has height, length, and width, but has these three dimensions in a specific relationship -- they are all equal. Distribute a handful of cubes to each student and ask how many cubes it would take to create one which is two cubes in height. After some experimenting, students will find that a "two-cube" is comprised of eight cubes. Next ask how many cubes they would need to make a "three-cube." They will find that it takes 27.

Tell students that you wish to find a way to calculate the exact number of cubes it would take to construct a cube of any size (an "N cube"). To facilitate, have them complete a table similar to that shown.

Height	Width	Length	Total cubes
1	1	1	1
2	2	2	8
3	3	3	27
4	4	4	64

Using the sugar cubes to construct the shapes, have each student create a formula (equation) for finding the total cubic units in a cube of any size. Students may generate formulas such as:
If n = height of cube, then n x n x n = total cubes OR n^3 = total cubes.

The underlying purpose is for students to discover that an equation can simply be a means of expressing a number pattern.

1 x 1 x 1 = 1   2 x 2 x 2 = 8   3 x 3 x 3 = ???

Hands On, Inc
2121 Rebild Drive
Solvang, CA 93463

Algebra

| 29 | Creates equations which reflect number patterns |

Time for My Formula!
Grade Level: Upper

MATERIALS: Various resource books -- almanacs, encyclopedias, conversion charts, etc.

ORGANIZATION: Individually or teams of two

PROCEDURE: This lesson teaches students to create equations by converting time measurements of seconds to minutes, to hours, to weeks, and to years.

Students have probably done simple time conversions such as how many hours in a week and how many minutes until school is over. In this activity, students will extend these simple tasks into creative formulas which will work for many scenarios. These formulas will have to take into account the changes of calendar days, holidays, weekends, etc.

Begin by giving a demonstration which will figure the total number of minutes until the end of the school year. We'll work from October 14 and imagine that the school year ends on June 21. One can proceed in a number of ways -- figure the total number of days and work backwards from this or figure the number of minutes in a 30, 31 and 28 day month and work from this. Students will come up with many other ideas as well and you should incorporate them into your explanation.

As you work through this problem with your students, interject questions such as, "How could this equation be changed to show the number of seconds in the school year?" or "How could you figure the actual number of minutes to be spent at school between now and summer?" Keep them thinking about how formulas can be changed to incorporate new stipulations. The emphasis of this lesson is in having students create a formula or equation which solves their problem. The correct answer is secondary.

Once you've worked through one problem with the students, let them create an equation for their own time frame. Have them share their results with the class.

Hands On, Inc
2121 Rebild Drive
Solvang, CA 93463

Algebra

29	Creates equations which reflect number patterns

What's it Worth to You
Grade Level: Upper

MATERIALS: Graph paper

ORGANIZATION: Individually

PROCEDURE: Recognizing patterns is an important skill in math. Algebra takes a pattern to the next step -- to express that pattern as an equation or formula. In this activity students will be working on developing this skill.

Tell students that they are going to be paid for being students. The pay will be represented by squares on a sheet of graph paper and everyone will start with the same amount of pay, 100 squares.

Distribute graph paper and have students cut out a block, 10 by 10 to represent their pay. Next, inform them that after one week of successful performance as a student, they will get an extra 15% of their base pay. Ask how they could represent this in terms of squares from the graph paper by cutting out an additional 15 squares for each week. Have them cut these out and place them next to their block of 100.

Ask how many extra squares they will have acquired after two succesful weeks of being a student and have them cut out a separate strip to represent this number. They should figure that if they get fifteen squares every week, then after two weeks they will get 2 x 15.

The next step is to have students create a table and ask how they could express this in an equation. They might assign the total number of squares the variable s and let w be the variable for the number of weeks. The constant is the number 15 (15% raise). The formula should be $s = (w \times 15) + 100$. To check the formula ask them how many extra squares they would have after 32 weeks. Have one student check by using a calculator and another check by cutting graph squares.

Week #	15% increase	total pay
1	15	115
2	15	130
3	15	145
4	15	160
5	15	175
6	15	190
7	15	205
8	15	220

Hands On, Inc
2121 Rebild Drive
Solvang, CA 93463

Algebra

| 29 | Creates equations which reflect number patterns |

Watch Out for the Algebug
Grade Level: Upper

MATERIALS: Beans

ORGANIZATION: Individually or in teams

PROCEDURE: One of the skills in algebra which students should become familiar with is how to represent a relationship using an equation, thus yielding a formula. To do this activity students should have knowledge of exponential multiplication.

Ask your students to pretend that they are medical scientists studying a new flu virus known as the AlgeBug. It is a new strain but behaves somewhat similar to all one celled life-forms, reproducing by cell division. If doctors can predict its growth rate, they will be able to accurately prescribe the dosage required to kill the virus.

Hand out 50 or so beans to each group. Have them arrange some beans to represent the following data: after one year one cell has split and become two virus cells; after two years these two cells have split and created four cells; after three years there are eight cells; and after four years there is a colony of 16 viral cells. Ask how many virus cells will be present after five years (32).

Next, ask them to explain any patterns they may have discovered (each successive year has doubled the number of the previous year). Divide them into cooperative groups of four to see if by manipulating the beans, they can create a formula which can predict the number of cells present after any number of years. One sample solution might be: 2 to the power of the desired year.

 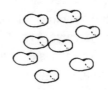 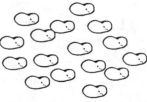

After 1 year = 2^1 After 2 years = 2^2 After 3 years = 2^3 After 4 years = 2^4

Algebra

| 30 | Creates solutions sets for equations using Venn diagrams, unions, intersections, etc. |

Ring Around the Answer
Grade Level: Upper

MATERIALS: Several teacher made two-step word problems, beans, paper clips, or some other multiple item for students to manipulate, string, small cards for labeling

ORGANIZATION: Cooperative groups of three or four

PROCEDURE: Venn Diagrams provide a highly visible and concrete experience in solving equations. In this activity, students will construct Venns using string (for the circles) and beans or paper clips to represent numbers in a variety of sample word problems (below).

Divide the class into groups of three or four and supply each group with 200 or so beans or clips. Hand out slips of paper with problem 1 and have them create a Venn at their table which represents the given numbers.

1) One hundred students watched a new T.V. program. All 100 had at least one reaction. If 48 students laughed and 72 students cried, how many students both laughed and cried?

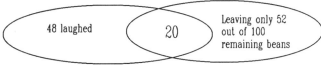

Therefore, you need 20 more to make 72, so 20 both laughed and cried

Explain to students that the question of how many <u>only</u> laughed or how many <u>only</u> cried is not addressed. See if students can write an equation which demonstrates this procedure: (48 + 72) -100 = x.

2) Forty-eight students clapped at the end of their T.V. program. 40 students clapped softly for their program and 1/2 of them clapped loudly. How many students clapped both loudly and softly.

3) Two hundred students viewed a horror movie and had these reactions: shock, amusement, or disgust. If 44 students were only shocked, 60 were only amused and 54 were only disgusted, 50 were both shocked and amused, 36 were both shocked and disgusted, and 24 were both amused and disgusted, how many students had all three reactions?

4) Each of 350 students viewed at least one of the following T.V. programs: Volcanoes, Canyons, and Mountains. Of the students who only saw one T.V. program, 90 saw only the Volcanoes show, 75 saw only the Canyons show, and 101 saw only the Mountains show. Of the students who saw more than one show 108 saw both the Volcanoes and the Canyons shows, 2/3 as many saw both the Volcanoes and Mountains shows, and 76 less than twice the Volcanoes and Mountains viewers saw both the Canyons and Mountains. How many students watched all three T.V. programs?

Hands On, Inc
2121 Rebild Drive
Solvang, CA 93463

Algebra

| 30 | Creates solutions sets for equations using Venn diagrams, unions, intersections, etc. |

Relating the Unrelated
Grade Level: Upper

MATERIALS: Pieces of string, blank cards, markers

ORGANIZATION: Cooperative groups of three or four

PROCEDURE: An understanding of terminology such as "variable," "set," "solution set," and "intersection" are important to a basic understanding of algebra. Venn diagrams provide a visual and tactile means for showing relationships between sets. In this activity, students will create Venn diagrams (string circles) and will create statements which require adjustments to their diagrams.

Divide the class into cooperative groups and give each group a stack of cards. Have them begin by writing a list of various types of cars (skateboards, music groups, or whatever else is "in"). Place all of these into one Venn diagram. From this information, have them create a statement which will divide the items listed into subgroups. Divide the cards so there are now two Venns. Next, have students figure out a common characteristic shared by one or two members in each Venn circle, thus creating an intersection. Have group members continue to experiment with forming intersections based upon shared attributes.

Eventually, students should write information in the form of set notation. As each group shares their information, emphasize terminology such as UNION, INTERSECTION, MEMBER, VARIABLE, SOLUTION SET, etc.

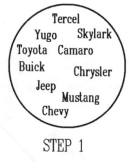
STEP 1

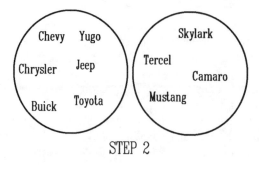
STEP 2

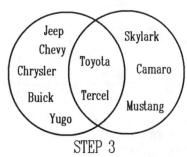
STEP 3

Algebra

| 30 | Creates solutions sets for equations using Venn diagrams, unions, intersections, etc. |

Varied Ads = Variables
Grade Level: Upper

MATERIALS: Many different magazines which would appeal to different audiences such as Sports Illustrated, Seventeen, Popular Mechanics, etc.

ORGANIZATION: Teams of two, three, or four students

PROCEDURE: Variables can present themselves in a number of formats. In magazines, a variable might be the type of advertisement used to appeal to particular clientele. In this activity, students will use the variables of advertisements to create Venn Diagrams.

Divide the class into groups of two, three, or four and give each group two or three different magazines. Their task is to cut advertising pictures from the first fifteen pages of each magazine. Once they have done this, have them create a Venn "collage" of the intersections of advertisements they found.

We also suggest you extend this lesson to have the students write a pargraph which explains why particular advertisements might be shared by two or three magazines (intersection) and why some advertisements might be specific to only one type of magazine.

Have students display their work and discuss the idea that the intersection in each collage might be considered to be "solution set."

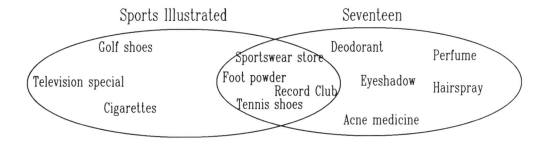

Hands On, Inc
2121 Rebild Drive
Solvang, CA 93463

Algebra

| 30 | Creates solutions sets for equations using Venn diagrams, unions, intersections, etc. |

Equating the Intersections
Grade Level: Upper

MATERIALS: Index cards with the numbers 0 – 10, pieces of string for Venn circles

ORGANIZATION: Cooperative groups of three or four students

PROCEDURE: In this activity, students will challenge their group partners to create greater than and less than equations by placing sets of numbers into Venn circles.

The teacher should begin by drawing two circles on the board (as shown) and placing numbers in each circle as well as in the intersection of the circles. The numbers must be in order (cannot be placed randomly). Students must also begin by making a statement such as, "for all positive numbers less than 6." This is necessary to limit the effects of the < or > symbols and to eliminate the need for the "..." after the highest number in the solution set.

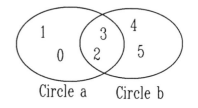

Ask students if they can create an equation which has the variables shown in circle a (0, 1, 2, 3). Some sample responses might be: n + 0 < 4 OR 3 – n < 4. The next step is to create an equation for circle b: 5 – n < 4 OR 3 + n > 4. You will find that students will have difficulty in generating these equations in that sometimes the equations seem to defy logic.

Once students understand the concept, divide them into cooperative groups of three or four. Give each group a set of number cards and strings for their Venn circles and have one student at a time pose problems to the other group members. The teacher should circulate among the groups to stimulate discussion and to view student responses.

As an extension, you may wish to use negative numbers as well as positive or use multiples of numbers (i.e. 3, 6, 9, 12 ...) so students can work with equations with multiplication and division.

Algebra

| 31 | Creates word problems based on given equations |

It Adds Up to My Family
Grade Level: Upper

MATERIALS: Students bring pictures or drawings of their families

ORGANIZATION: Individually after being explained as a whole class

PROCEDURE: In this activity, students will be using the members of their families to create equations and to present word problems. This will give them extra practice with using greater than, less than, and equal to signs.

On the day before the lesson, ask each student to bring several pictures of their brothers, sisters, and parents or other family members. From these pictures, the students should create different types of word problems. These can be done by age, by height, by weight, etc. A sample problem is shown below. As students become more sophisticated, they can begin to use equations involving square roots or parentheses and brackets.

Once students have created an equation that uses all members of the family, they should tape the pictures on a piece of paper and display them on a bulletin board. Have each student come forward to explain their equation to the whole class. An alternative is to have them write the information as a paragraph to share with the whole class.

---

d = dad   m = mom   s = sister   b = brother   y = younger brother   i = me

Dad is forty and is two years older than mom, and mom is twice as old as my sister, if my younger brother is eight years younger than I am, and if I am eight years younger than my sister, how old is my younger brother?

$$m = (40 - 2)$$
$$s = m \div 2$$
$$i = s - 8$$
$$y = i - 8$$

$$[(40 - 2) \div 2] - 16 = y$$

Hands On, Inc
2121 Rebild Drive
Solvang, CA 93463

Algebra

| 31 | Creates word problems based on given equations |

Records With a Club!
Grade Level: Upper

MATERIALS: Flyers and advertisement for record/tape clubs which appear in magazines

ORGANIZATION: Individually or small groups of students

PROCEDURE: Most students are familiar with the record club advertisements which appear in many magazines -- the headline says, "Buy one, get ten free." Most students also know that there is fine print in these advertisements. This activity has students writing an extended word problem to incorporate the information contained within one of these ads.

Give each student a copy of one of the record advertisements and have them read carefully, noting any extra costs or catches as well as the number of tapes, CD's or records they will receive. In such an ad which we analyzed, the following "facts" were gleaned.

1. Twelve cassettes for $.01.
2. $1.85 for shipping and handling on the first twelve cassettes.
3. Promise to buy eight more tapes in the next three years.
4. Receive nineteen mailings for tape offers each year.
5. If you don't want the selected item in each of the nineteen mailings, mail a card back.
6. If no card is mailed, the selected item will be sent to you at a cost of $9.98 per cassette + $1.85 shipping and handling.

From this information, have your students create a word problem which can be solved by classmates. A variety of possible queries might include: if no cards were ever returned, how much would the club cost over the three years? What would be the average price per cassette? What is the minimum cost of the club inlcuding postage of $.16 with each return mailing? Given current store prices, what is the difference in cost between 20 tapes with the club and 20 tapes with a store? etc.

You may find that students will become very shrewd in their problems by addressing other issues such as sales tax and limitations of choice. Have students share their word problems with the entire class.

Hands On, Inc
2121 Rebild Drive
Solvang, CA 93463

Algebra

| 31 | Creates word problems based on given equations |

Equation Collage
Grade Level: Upper

MATERIALS: Magazines, newspapers, paper, paste, scissors

ORGANIZATION: Whole class

PROCEDURE: The object of this lesson is to create word problems from equations using pictures to clarify the word problem. The equations can be teacher generated or student generated.

At first you may want the whole class to work from one equation, later you may wish to display a list of equations and let them choose from the list.

The students should find pictures to put in a collage that reflect numbers in the equation. They will then write a word problem based on the pictures and equation. Once completed, students should exchange their collages and word problems and have classmates solve them.

A sample equation, collage and word problem representation might be:

$$3n + 4 - (2 + 3) = 5$$

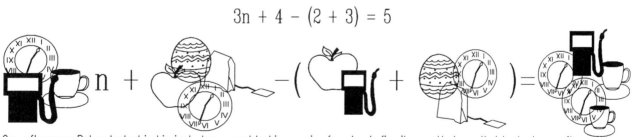

One afternoon, Ruben looked in his junk drawer and to his surprise found only five items. He knew that he had more items than that so he tried to remember what he had given away. He had given an apple and a miniature gas pump to Tony, and had given an old watch, an egg, and an herbal tea bag to Phil. Roger had given him an old watch, an apple, a painted egg, and an herbal tea bag. Originally, he remembered having more than one watch, miniature gas tank and cup. In the equation shown above, what must n equal to show what he started with in his junk drawer?

Hands On, Inc
2121 Rebild Drive
Solvang, CA 93463

Algebra

31	Creates word problems based on given equations

The Whys of Ups and Downs
Grade Level: Upper

MATERIALS: Teacher prepared stock lists (like the Disney information given below)

ORGANIZATION: Groups of two, three, or four students

PROCEDURE: The movement of stocks from day to day or week to week can provide a means of creating equations. This activity uses this movement to have students create stock exchange scenarios.

Divide the class into groups of two, three or four and give each group a different stock to follow. We suggest that the teacher prepare simulated stock movement for companies in which students may have an interest. For this example we will use Disney stock which has had movement of 131 5/8 to 125 1/8 to 121 1/2 to 140 1/4.

Given this information, the students should write an extended word problem in which they not only tell about the up and down price of the stock but also the reason why it moved up or down. A sample student response might be:

> When we first bought Disney, Inc., the price was 131 5/8, which really means we paid $131.63 for a share. Soon after we bought it, they released a new movie called, The Little Mermaid, Part II which cost several million dollars to make but was unsuccessful at the theaters, causing the stock to drop $6.50 per share. The company increased advertising for the film but it continued to fail at the box office. The next time we looked, it had dropped another 2.9%. Fortunately, there was a song in the movie which was recorded by an "in" rock group and sales on the Disney record skyrocketed, causing our stock to jump $18.75. What is our Disney stock worth now?

When students finish their work, they should check the computations and write an equation which will give the solution. Have them share their creations with the rest of the class for discussion and computation.

Hands On, Inc
2121 Rebild Drive
Solvang, CA 93463

Algebra

| 32 | Creates equations from information found in tables and graphs |

Making Converts
Grade Level: Upper

MATERIALS: Various resource books -- almanacs, atlases, encyclopedias, conversion charts, etc.

ORGANIZATION: Cooperative groups of four

PROCEDURE: This lesson provides a means to teach equations to students by having them convert minutes to seconds to hours, or inches to feet to yards by writing them as equations.

Begin by giving each group a different topic to work with. If you have six groups in your room, assign the following topics: liquid measurement, linear measurement, time, weight, metric distance, and volume.

Sample answers which you might ask students to provide might be:

- Liquid measurement: How many teaspoons of water would there be in a pool 50' long, 20' wide, and 8' deep?
- Linear measurement: How many inches between Los Angeles and Chicago if you take the shortest route on interstate highways?
- Time: How many seconds in a leap year?
- Weight: How many ounces would there be in 2 tons of gold?
- Metric distance: How many millimeters around the equator?
- Volume: How many cubic inches in the Sears Tower Building?

Students will probably need to work out the solution first and then come back to the writing of an equation. Eventually, they should write an equation which can be given to other groups to use without having to go through the reasearch of converting.

Have each group share their process with the entire class.

Hands On, Inc
2121 Rebild Drive
Solvang, CA 93463

Algebra

| 32 | Creates equations from information found in tables and graphs |

Sometimes You Win...
Grade Level: Upper

MATERIALS: Sports section data from won-lost records, box scores, etc.

ORGANIZATION: Cooperative groups of four or individually

PROCEDURE: Many students are inherently interested in sports. In this activity, students will get a list of team standings, box scores, or other sports statistics in which numbers have been assigned variables. They will have to go through and figure out the unknowns.

Below, we have used examples of a baseball box score and a set of football standings (football works well because all teams have generally played the same number of games at any given time).

Begin by showing a copy of the baseball box score shown below. Tell the students that there are three unknowns: x, y, and z. Give students time to see if they can solve for the unknowns, but it is important that they should be able to explain the process they used to the whole class. Eventually, you will want them to write the solution method as an equation.

Divide the class into cooperative groups and give each group a copy of the sports statistics page. Have them select one or two of the game reports and have them assign variables to one, two, or three numbers. They then need to create an equation which will allow classmates to solve for the unknown. Circulate the student generated problems around the room for each cooperative group to solve.

	At Bats	Hits	RBI	Runs
Henry	5	1	0	1
Smith	5	3	0	1
O'Meara	4	x	0	0
Lenger	z	1	2	y
Brisby	4	2	0	0
Perez	4	0	0	0
Hernandez	3	0	0	1
Long	1	0	1	0
Pfau	3	1	0	1
Redkey	2	1	1	0
Johnson	3	2	2	0
Ripple	1	0	0	0
Ibsen	1	0	0	0
TOTALS	39	13	6	6

Hands On, Inc
2121 Rebild Drive
Solvang, CA 93463

Algebra

| 32 | Creates equations from information found in tables and graphs |

A Yen for Formulas
Grade Level: Upper

MATERIALS: Precut rectangles of different colored construction paper about the size of a dollar bill, markers, reference books with monetary equivalents

ORGANIZATION: Teams of two students

PROCEDURE: To begin this lesson, the students must first research the equivalent value of units of money of different countries. They must then record these equivalents on a chart or make facsimiles to share with one another.

Give students a selection of colored rectangles and have them choose a color for each country's monetary units. For example, black might represent U.S. dollars; red, Japanese yen; and white, English pounds, etc. Unknowns in each case would be the value in U.S. dollars in another country's money.

Divide students into groups of two and have them lay out the rectangles. Have student one pose a question such as, "What is the formula for converting yen to dollars?" Student two should then figure the equivalent and write the equation.

As an extension, you might set up a store and allow students to take turns buying and selling items with the different types of money.

Mexico	2228 pesos	1 U.S. Dollar
Japan	125 yen	1 U.S. Dollar
China	4 yuan	1 U.S. Dollar
France	5 francs	1 U.S. Dollar
Germany	2 marks	1 U.S. Dollar

Hands On, Inc
2121 Rebild Drive
Solvang, CA 93463

Algebra

| 32 | Creates equations from information found in tables and graphs |

Three Coins in a Pocket
Grade Level: Upper

MATERIALS: Coins (either those in the students' possession or packets of coins given to each student -- Appendix D)

ORGANIZATION: Cooperative groups of four

PROCEDURE: This activity uses the random number of coins found in students' pockets to generate numbers for use in creating equations. You may find that you will need to give students "paper coins" from the set given in appendix D.

Divide the class into groups of four and have students place all of the coins in their possession on the desk in front of them. Each student should count the amount, write it on a slip of paper and have each group member initial the paper in hopes of eliminating arguments when the coins are redistributed at the end of class.

From this conglomeration of coins, have each student generate three equations. First, an equation that will total the monetary amount; second, an equation that will provide a formula for figuring the value of any set of coins; and third, an equation that will figure the average amount of money possessed by each group member. Sample equations are shown below.

Student 1 = 3 pennies, 2 dimes, 2 quarters
Student 2 = 1 penny, 4 nickles, 2 dimes, 3 quarters
Student 3 = 2 nickles, 2 dimes
Student 4 = 5 dimes, 6 quarters

Equation 1: $4(1) + 6(5) + 11(10) + 11(25) = \text{total}$
Equation 2: $1(p) + 5(n) + 10(d) + 25(q) + 50(h) = \text{total}$
Equation 3: $[\, 1(p) + 5(n) + 10(d) + 25(q) + 50(h)\,] \div 4 = \text{average}$

As an extension, you might have students extend their formulas to include dollar amounts or to include larger numbers of students; perhaps even a whole class net worth equation.

Hands On, Inc
2121 Rebild Drive
Solvang, CA 93463

Algebra

| 33 | Creates a simplified expression of an equation |

Bringing a Monopoly Down to Size
Grade Level: Upper

MATERIALS: Several sets of the game "Monopoly" brought by students

ORGANIZATION: Cooperative groups of four

PROCEDURE: Many games that students play at home are excellent for practicing math in a real-life situation. This activity uses the game Monopoly as a means of simplifying complex equations.

Divide the class into teams of four with each group having a Monopoly gameboard. It is suggested that the teacher direct the money distribution and rule explanation as many students play by their own rules.

Give students twenty minutes or so to begin the game. The one added feature is that each time they make a move or conduct a transaction, they must make a note of the cost, change made, etc. This is the information that they will be eventually simplifying in an equation.

You will find that Monopoly is a wonderful game for teaching math skills if it is played as described in the directions -- i.e. auctioning of properties, trading real estate, mortgaging, and improving property with houses and hotels.

After twenty minutes, have students stop the game and ask each student to write an equation which describes the transactions which have taken place.

$1,500 − $60 − $100 − $150 − $180 − $26 − $300 + $6 − $60 + $200
initial amount buys Baltic Ave. buys Connecticut Ave. buys Electric Co. buys Tenessee Ave. rent on Kentucky Ave. buys Pacific Ave. rent on Baltic buys Mediteranean Ave. passes GO

To simplify: ($1,500 + $200 + $6) − ($60 + $100 + $150 + $180 + $300 + $60) = CASH
1,706 − $850 = CASH

Hands On, Inc
2121 Rebild Drive
Solvang, CA 93463

Algebra

| 33 | Creates a simplified expression of an equation |

Bean Counters Take Heart
Grade Level: Upper

MATERIALS: An assortment of beans

ORGANIZATION: Individually or in pairs

PROCEDURE: In this activity students will practice their ability to write and then simplify an expression. This is important in algebra because it prevents the occurrence of long or awkward equations, making it easier to see relationships.

Pass a handful of beans to students and have them write down an equation for the contents of their bean pile. For example, if the pile has five pintos, twelve navy, eight kidney, and three lima beans, they would write $5p + 12n + 8k + 3l = T(total)$

Explain that they will be creating a "bean currency" for their pile. The bean with the <u>highest value</u> is the bean of which they have the <u>least amount</u>. In the set given above, "l" would have the greatest value. The beans with the least amount of value would be n; this will be the "penny" of the pile.

Next, roll a die three times to give each of the other types of beans a value relative to "n." If you roll a six, three and five, then the lima would be worth 6n (six times greater than n), p = 5n, and k = 3n. Next, have your students express their equations in terms of n: $5(5n) + 12(n) + 8(3n) + 3(6n)$ and simplify further if possible.

Once they are familiar with this process, redistribute the beans and change the values.

Algebra

| 33 | Creates a simplified expression of an equation |

Have You Got Five Fizzles for a Klatch?
Grade Level: Upper

MATERIALS: Magazines/catalogs, scissors, glue/tape, etc.

ORGANIZATION: Individually or in pairs

PROCEDURE: In this activity students will practice their ability to express and simplify values using variables.

Explain to your students that they are going to visit Algebropolis on the planet of Bleemur. They are going to be commercial ambassadors from earth and will need to create a catalog of goods they are going to sell to the aliens. Have them look through the catalogs, cut out pictures, and then paste one picture per page to create a new catalog; each catalog should have at least 25 items for sale.

They may charge as much as they like for any item as long as it is under $500,000. The larger the number, the more trouble they will have with exchanging the dollar price into fizzles, the currency in Algebropolis. The price, should be a multiple of the original price (rounded to the nearest dollar). They should make an attractive cover and give a brief description of their products.

Tell them that they must now list the price of each item in fizzles. This currency converts as follows:
    1 FIZZLE = $1.50   1 DROON = 6 blaubs, 10 fizzles
    1 BLAUB = 15 fizzles   1 KLATCH = 5 droons
    MEGA is used to indicate a quantity multiplied by 25 (a MEGA-FIZZLE = 25 fizzles)

Your students' job will be to express all the prices in their catalog in Bleemurian terms. For example: a $250 stereo might be multiplied by 3 to a cost of $750. Since f (fizzle) = $1.50, this stereo would cost 500 fizzles, or s = 500f. Unfortunately, Algebropolis demands that all prices be listed in simplest terms. Since a droon equals 100f, then the cost could be 5d; this lesson is to gain practice simplifying expressions, have your students go through all the steps of translating prices from dollars to fizzles to the appropriate larger denominations.

Hands On, Inc
2121 Rebild Drive
Solvang, CA 93463

Algebra

| 33 | Creates a simplified expression of an equation |

Thanks, but No Thanks!
Grade Level: Upper

MATERIALS: No special materials are necessary

ORGANIZATION: Individually

PROCEDURE: In this activity, students will use the famous Christmas song, The Twelve Days of Christmas as a source of information for an extremely long equation which can then be simplified.

Most, if not all, of your students will be familiar with the song, but you may want to review the items used with each of number. They are...

first	a partridge in a pear tree (a)
second	two turtle doves (b)
third	three french hens (c)
fourth	four calling birds (d)
fifth	five golden rings (e)
sixth	six geese a laying (f)
seventh	seven swans a swimming (g)
eighth	eight maids a milking (h)
ninth	nine ladies dancing (i)
tenth	ten lords a leaping (j)
eleventh	eleven pipers piping (k)
twelfth	twelve drummers drumming (l)

The next step is to give each item a variable and then begin writing the equation for each verse. The final result will be...

$$12a + 22b + 30c + 36d + 40e + 42f + 42g + 40h + 36i + 30j + 22k + 12l = \text{song items}$$

You might want to discuss why the pattern of numbers is created.

Hands On, Inc
2121 Rebild Drive
Solvang, CA 93463

Appendix A   Algebra

H	A	N	D	O
		🖐		

Hands On, Inc
2121 Rebild Drive
Solvang, CA 93463

Appendix B Algebra

Rollin' Around

STARTα

Appendix C   Algebra

Cryptograms

	CODE 1	CODE 2
A	1	2
B	2	4
C	3	6
D	4	8
E	5	10
F	6	12
G	7	14
H	8	16
I	9	18
J	10	20
K	11	22
L	12	24
M	13	26
N	14	25
O	15	23
P	16	21
Q	17	19
R	18	17
S	19	15
T	20	13
U	21	11
V	22	9
W	23	7
X	24	5
Y	25	3
Z	26	1

Hands On, Inc
2121 Rebild Drive
Solvang, CA 93463

Appendix D

Algebra

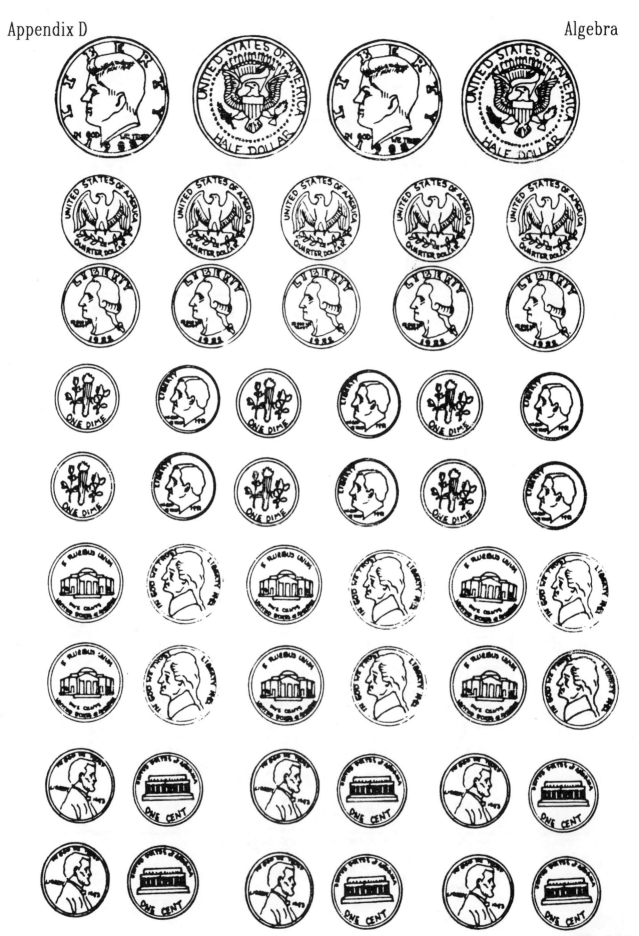

Appendix E — Algebra

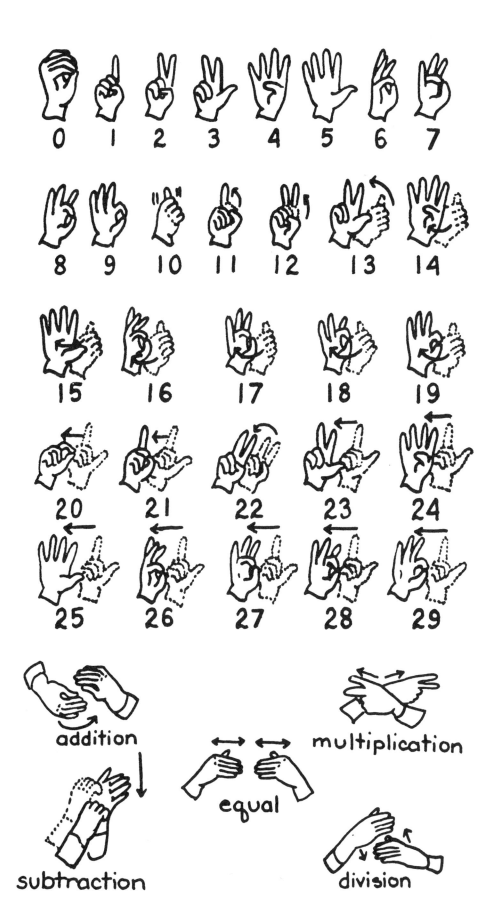

181

Appendix F   Algebra

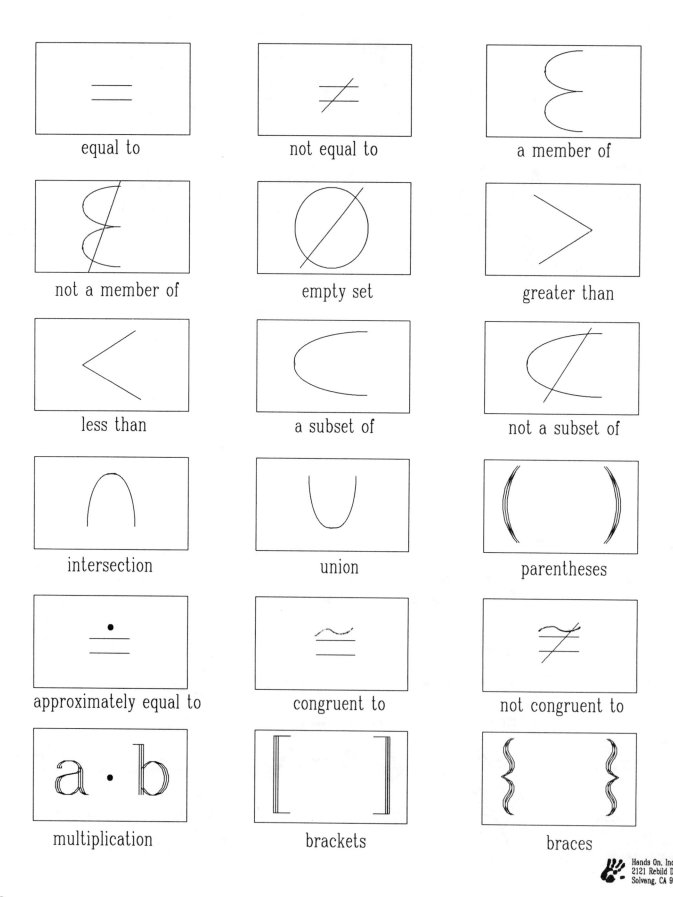

Appendix F (page 2) Algebra

Question Cards

| 3 □ III | 1−0 □ 1×0 | Tuesday □ Days of the Week |

| Abe Lincoln □ Living Presidents | {x,y,z} □ {z,x,y} | January □ Days of the Week |

| 5 □ 4 □ 3 | 0 □ 1 □ 3 | U = {1,2,3,4,5}
T = {2,4,5}
T □ U |

| U = {1,2,3,4,5,6,7}
T = {0,1,7}
U □ T | S = {1,3,5,7,9}
T = {0,3,6,9}
S □ T | S = {letters in word "rice"}
T = {Letters in word "nice"}
S □ T |

| a = 9 b = 7
ab = □ | □ 2−3 □ +7=13 | 28949 □ 30000 |

Hands On, Inc
2121 Rebild Drive
Solvang, CA 93463

Appendix G   Algebra

1. Juan, Sharon, and Andy were playing a game of darts. Juan made a score of 3 points, Sharon scored two times as many points as Juan. Andy scored one more than Sharon. How may points did Andy score?

2. The Eighth grade students were helping the kindergarten students in the computer lab. There were five eighth grade students helping. There were four times as many kindergarteners as eighth graders. How many students were in the computer lab?

3. Tony agreed to bake 125 cookies for the class party. He baked 42 cookies on Wednesday and 36 cookies on Thursday. How many more cookies were left to bake?

4. Scott and Petti collect baseball cards. Scott has six 1935 baseball cards and four times as many 1975 cards. Petti has 42 cards all together. How many more does Petti have than Scott?

5. The fourth grade won 13 soccer games. The fifth graders won 18 soccer games. The sixth graders won 12 more soccer games than the fourth graders, how many soccer games were won?

6. Ling put rabbits in three cages. Natalie and Ron put rabbits in two times as many cages as Ling. There are ten cages. Have they filled all of the cages? If not, how many cages are empty?

SOLUTIONS:

1.	(2 x 3) + 1 = a	4.	42 − (4 x 6) = b
2.	(4 x 5) + 5 = s	5.	(13 + 18) + (13 + 12) = g
3.	(125 − 42) − 36 = c	6.	10 − (2 x 3) = c

Hands On, Inc
2121 Rebild Drive
Solvang, CA 93463

Appendix H   Algebra

EQUATIOCARD

			=	
			<	
			>	

Hands On, Inc
2121 Rebild Drive
Solvang, CA 93463

Appendix I — Algebra

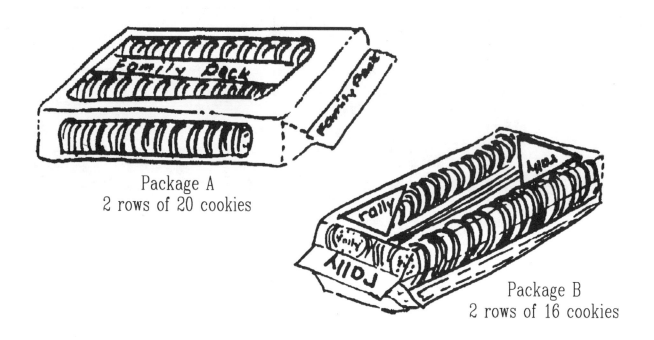

Package A
2 rows of 20 cookies

Package B
2 rows of 16 cookies

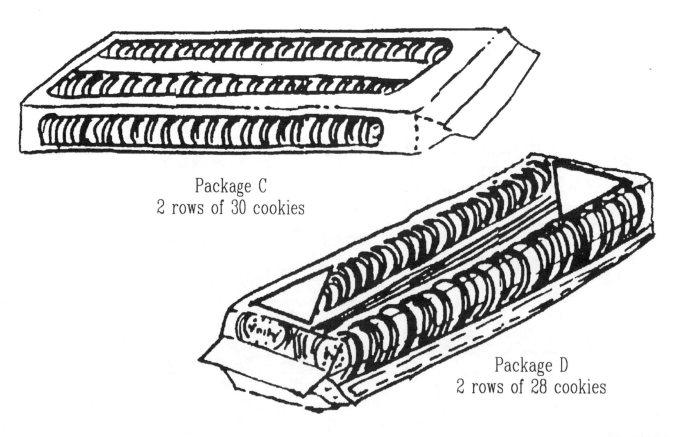

Package C
2 rows of 30 cookies

Package D
2 rows of 28 cookies

Hands On, Inc
2121 Rebild Drive
Solvang, CA 93463

HANDS ON, INC.
2121 Rebild Drive
Solvang, CA 93463

ORDER FORM
Please Print

Bill To: _____

Ship to: _____

_____

_____

Phone: _____

Attention: _____

Purchase Order No:	Payment enclosed ☐	Please Bill ☐	Date:

Quantity	H.O. Number	Book Title	Unit Price	Total
	101	Hands On Statistics, Probability & Graphing	$15.95	
	102	Hands On Measurement	$15.95	
	103	Hands On Logic	$15.95	
	104	Hands on Geometry	$15.95	
	105	Hands on Algebra (2/90)	$15.95	

Check one: ☐ Mastercard ☐ VISA

Account Number _____

Signature: _____
A really essential component!

Interbank #: ☐☐☐☐   Expiration Date: ☐☐ ☐☐
(Above your name on the card)   month year

Authorization Number _____

All Orders are shipped U.P.S.
Shipping Charges to Alaska, Hawaii, APO, and
Canada will be higher. Write for confirmation.

Total for Materials	
Sales Tax (CA residents)	
Shipping and Handling $2.50 for first book $1.00 each additional book	
TOTAL	

Order by Phone
24 Hrs/day
7 days/wk
(805) 688-8868

THANK YOU FOR YOUR ORDER!!

HANDS ON, INC.
2121 Rebild Drive
Solvang, CA 93463

ORDER FORM
Please Print

Bill To: _____   Ship to: _____

_____   _____

Phone: _____   Attention: _____

Purchase Order No:	Payment enclosed ☐	Please Bill ☐	Date:

Quantity	H.O. Number	Book Title	Unit Price	Total
	101	Hands On Statistics, Probability & Graphing	$15.95	
	102	Hands On Measurement	$15.95	
	103	Hands On Logic	$15.95	
	104	Hands on Geometry	$15.95	
	105	Hands on Algebra (2/90)	$15.95	

Check one: ☐ Mastercard   ☐ VISA

Account Number _____

Signature: _____
A really essential component!

Interbank #: ☐☐☐☐   Expiration Date: ☐☐ ☐☐
(Above your name on the card)   month year

Authorization Number _____

All Orders are shipped U.P.S.
Shipping Charges to Alaska, Hawaii, APO, and
Canada will be higher. Write for confirmation.

Total for Materials	
Sales Tax (CA residents)	
Shipping and Handling $2.50 for first book $1.00 each additional book	
TOTAL	

Order by Phone
24 Hrs/day
7 days/wk
(805) 688-8868

THANK YOU FOR YOUR ORDER!!